Energy Access and Forced Migration

This edited collection brings together a selection of expert authors and draws on a wide range of case studies, geographies, and perspectives to explore the links between forced migration and energy access.

This book addresses the paucity of academic study on how energy is delivered to the millions of people currently forcibly displaced. The contributions throughout assess the current energy governance regimes, models of delivery, and innovative solutions that are dictating how energy is – and can be – provided to those who have been forced to move away from their homes. By bringing together author-teams of practitioners, academics, businesses, and policy makers, this collection encourages interdisciplinary dialogue about the best way of approaching energy provision for the forcibly displaced.

This book will be of great interest to students and scholars of energy access and policy, environmental justice and equity, and migration and refugee studies.

Owen Grafham is Department Manager of the Energy, Environment and Resources Department at The Royal Institute of International Affairs, Chatham House, UK.

Routledge Studies in Energy Transitions

Considerable interest exists today in energy transitions. Whether one looks at diverse efforts to decarbonize, or strategies to improve the access levels, security and innovation in energy systems, one finds that change in energy systems is a prime priority.

Routledge Studies in Energy Transitions aims to advance the thinking which underlies these efforts. The series connects distinct lines of inquiry from planning and policy, engineering and the natural sciences, history of technology, STS, and management. In doing so, it provides primary references that function like a set of international, technical meetings. Single and co-authored monographs are welcome, as well as edited volumes relating to themes, like resilience and system risk.

Titles in this series include:

Energy Access and Forced Migration
Edited by Owen Grafham

For more information about this series, please visit: www.routledge.com/Routledge-Studies-in-Energy-Transitions/book-series/RSENT

Energy Access and Forced Migration

Edited by Owen Grafham

Routledge
Taylor & Francis Group
LONDON AND NEW YORK

earthscan
from Routledge

First published 2020
by Routledge
2 Park Square, Milton Park, Abingdon, Oxon OX14 4RN

and by Routledge
605 Third Avenue, New York, NY 10017

First issued in paperback 2021

Routledge is an imprint of the Taylor & Francis Group, an informa business

Publisher's Note
The publisher has gone to great lengths to ensure the quality of this reprint but points out that some imperfections in the original copies may be apparent.

British Library Cataloguing-in-Publication Data
A catalogue record for this book is available from the British Library

Library of Congress Cataloging-in-Publication Data
A catalog record for this book has been requested

ISBN 13: 978-1-03-208213-4 (pbk)
ISBN 13: 978-1-138-54338-6 (hbk)

Typeset in Times New Roman
by Apex CoVantage, LLC

Contents

Figures

Tables

Contributors

Raffaella Bellanca, Energy for Food Security coordinator at WFP

Raffaella is an Access to Energy specialist focused on energy delivery models and the development of sustainable value chains that meet households, institutional, commercial and humanitarian needs for cooking, lighting, cooling and productive uses. She has co-authored several papers and a book on the subject. She entered the development sector as executive director of the energy practitioners network, HEDON, publisher of the peer reviewed journal Boiling Point. She has worked in the field, Haiti and Mali, as well as in London.

Raffaella has worked in the energy sector for more than twenty years, starting from simulating combustion processes in power plants (for ENEL SPA) and car engines. She experienced entrepreneurship first hand by co-founding a university spin-off clean-tech company, incubated by a technology centre in Sweden.

She holds a PhD in Combustion Physics, MSc in Environmental Physics and a Master in Communication for Development.

Iwona Bisaga, University College London

Iwona Bisaga is Postgraduate Researcher at the University College London and Research Manager at BBOXX – an off-grid solar and clean cooking company operating across Sub-Saharan Africa and Asia. Her research focuses on energy policy, decentralised energy access in rural, urban and humanitarian contexts, as well as infrastructure provision and environmental management in informal settlements.

Anoushka Boodhnais

Anoushka Boodhnais a market systems development specialist with 15 years professional experience. She has worked with development organisations, private sector and SMEs, and spent more than 6 years living and working across East Africa, West Africa, Latin America and Asia. In the past, she worked in retail in the private sector.

Tamsin Bradley, University of Portsmouth

Tamsin Bradley is Professor of International Development Studies at the University of Portsmouth, UK. She is a social anthropologist who has researched, taught and consulted on violence against women and girls in different development contexts in South Asia and Africa for 17 years. She is the author of seven books and more than 20 peer-reviewed articles and has secured research grants and consultancy contracts from DFID, ESRC, EU, and the British Academy. Tamsin holds a PhD in Social Anthropology, Gender and International Development from the School of Oriental and African Studies, University of London, UK. orcid.org/0000-0003-3472-5723

Elena Gaura, Coventry University

Professor Elena Gaura received her PhD in Intelligent Sensor Systems in 2000 (Coventry University). She was appointed as inaugural director of the Coventry University's Cogent Computing Applied Research Centre in 2006 (a position she held until her appointment in 2013 as Associate Dean for Research) and was awarded a Professorship in Pervasive Computing in 2009. Over the course of her career, Elena has accrued a sturdy academic reputation in the area of MEMS based smart sensing systems in general and wireless sensor networks (WSNs)/Internet of Things (IoT) in particular. Her work is for and with the industry and society, with current research focus on community energy provision for impoverished groups, globally. Elena's research draws from advances in renewable energy, control and the IoT to bring forth socially acceptable, sustainable energy systems designs that are fit for purpose and context and deliver social empowerment.

Lama Gharaibeh, Norwegian Refugee Council (NRC)

Lama Gharaibeh works on supporting the Syrian Refugee Crisis in Jordan under the shelter and Energy Programs, developing solutions to refugees and host communities by linking the energy aspects on the program activity in Shelters, schools and Youth empowerment and Advocating for the humanitarian sector to take a role in adapting energy responses in the programs and operations; currently Lama is the Renewable Energy Technical Officer in Jordan.

Owen Grafham, Chatham House

Owen Grafham joined Chatham House in May 2014. During his time in the Energy, Environment and Resources department, he has managed Chatham House's research and outreach on energy for displaced populations and the institute's work on energy use inside the humanitarian system. He is on the technical advisory board for a qualitative research project funded by Economic and Social Research (ESRC) at Edinburgh University, which focuses on the energy use of refugees, and an ESRC-funded project at Coventry University studying humanitarian energy for displaced populations in camps and informal settlements (HELP). Owen has a BA from the University of York and an MSc in African Politics from the School of Oriental and African Studies, London.

Diego Hakspiel, Sanivation

Diego Hakspiel is the Director of Partnerships at Sanivation; he leads the implementation and scale up of a waste to value sanitation system in Kakuma refugee camp and oversees partnership opportunities with the Kenyan government and humanitarian partners worldwide. Diego has over ten years working in the humanitarian sector for UNHCR, UNICEF, and UNODC and holds a master's degree from the London School of Economics in international development and humanitarian emergencies.

Mansoor Hamayun, BBOXX

Mansoor Hamayun is co-founder and Chief Executive Officer of BBOXX, a next generation utility powering growth and transforming lives. BBOXX uses smart, solar technology alongside pioneering financing techniques to improve energy access across Africa and the developing world. Born in 1988 in Pakistan, Mansoor was raised in Sweden and studied Electrical Engineering at Imperial College London. Whilst at university he founded the student charity, e.quinox, which brought electricity to six villages in Rwanda.

Elias Jourdi, Norwegian Refugee Council (NRC)

Elias Jourdi is a humanitarian professional with over 12 years of experience working on designing and implementing programs to support refugee, IDP, and returnee populations in emergency and recovery situations in Kenya, Ethiopia, Yemen, Jordan, and Lebanon. He works with the Norwegian Refugee Council (NRC) as the Global Shelter and Settlements Advisor and previously as the Shelter and Energy Specialist with NRC Jordan. He holds a MSc in Mechanical Engineering – Energy from the Lebanon University in Beirut.

Glada Lahn, Chatham House

Glada Lahn is a Senior Research Fellow in the Energy, Environment and Resources Department at Chatham House. Since joining Chatham House in 2004, Glada Lahn has led a range of international research projects, in relation to access to energy in developing countries, sustainable transitions in oil and gas-exporting economies, and how climate change and decarbonization affect the prospects and choices for developing country oil and gas producers.

Johanna Lehne, E3G

Johanna Lehne is a policy advisor at E3G in Brussels. Before joining E3G, she was a research associate in the Energy, Environment and Resources Department at Chatham House. Her research includes industrial decarbonization, the circular economy, low-carbon innovation, chokepoints in global food trade, resource use in China, and energy for displaced populations.

Katherine Liakos, IMC Consulting

Katherine Liakos is an international development professional with a technical focus on monitoring, evaluation, and learning. Katherine has worked on and managed multiple projects for donors such as DFID and USAID in Africa and Asia. Prior to this, she worked for an international development consultancy in Ethiopia, the European Commission's DG DEVCO, and nonprofit and civil associations in Peru. Katherine holds an MA in development studies and a BA in French and linguistics, both from the University of Leeds, UK.

Eva Mach, International Organization for Migration (IOM)

Eva Mach works as Environmental Sustainability Programme Officer at the Headquarters of the International Organization for Migration – the UN Migration Agency (IOM). In this capacity, she manages IOM's institutional Environmental Sustainability Programme, which aims to connect environmental sustainability principles and practices with migration governance and management. She also contributes to IOM's global policy work on migration, environment, and climate change, in particular on water and energy-related topics.

Jonathan Daniel Nixon, Coventry University

Jonathan Daniel Nixon is Senior Lecturer and MSc Renewable Energy Engineering course director at Coventry University. He gained his PhD in 2012 from Aston University whilst working on hybrid solar-biomass power plants, and, as a research fellow, he assisted in the development of pyrolysis plants for generating bio-fuels from agricultural residues in India. Recently, he has been working on improving the efficiency on wind turbines located in cold climates. He is currently the Co-I on a £1.1M EPSRC project, Humanitarian Engineering and Energy for Displacement (EP/P029531/1), developing smart-enabled energy interventions in refugee camps and displaced settlements for improved control and management of shared communal energy resources, including a micro-grid, advanced streetlights, and a community hall.

Shada Qahoush, Norwegian Refugee Council (NRC)

Shada works on humanitarian response and recovery programs, focusing on emergency shelter and housing solutions for people affected by displacement. She has previous experience in Jordan and Iraq and is currently the Shelter Programme Development Manager in NRC Myanmar.

Sarah Rosenberg-Jansen, University of Oxford

Sarah Rosenberg-Jansen is an expert adviser on humanitarian energy, with experience leading sustainable energy work with displaced communities in several countries, including Rwanda, Kenya, Burkina Faso, Jordan, and Nepal. Sarah researches renewable energy transitions and socio-technical networks at the University of Oxford and is a published academic as well as senior practitioner within the humanitarian energy sector. She has worked across a range of renewable energy and regulatory technical issues for Imperial College London, the Climate Policy Initiative, and the University of Cambridge. Her Twitter handle is: @SarahLRosenberg

Niraj Subedi, KFW

Niraj Subedi holds over 18 years of diverse experience as an academic, a design engineer, and a development expert. He since 2012 has been working as an energy sector specialist in KFW, the German Financial Cooperation in Nepal. With his interest, he is engaged in several research projects in the areas of renewable energy and energy resilience.

Long Seng To, Loughborough University

Long Seng To is an Engineering for Development Research Fellow funded by the Royal Academy of Engineering at Loughborough University. Her research tackles the challenge of providing access to affordable, reliable, sustainable, and modern energy for all in the context of increasing stresses and shocks, such as climate change, disasters, and conflicts. Her research focuses on enhancing community energy resilience using renewable energy in South Asia and Sub-Saharan Africa.

Mattia Vianello, Practical Action

Mattia Vianello is an energy access expert currently heading Practical Action in West Africa. His focus is on promoting energy access by unlocking market systems in development and humanitarian contexts. Mattia has more than 10 years of' experience in consultancy, research and project implementation across Europe, Africa, Central America and Asia. He holds an MSc in Science and Technology Policy and Management from Edinburgh University and a BA in Philosophy from Vita-Salute San Raffaele in Milan, Italy.

Acknowledgements

I am grateful to many of my current and former colleagues at Chatham House who have provided feedback on individual chapters and the overall framing behind this edited collection. Particular thanks to Glada Lahn who has devoted much time to making this collection better, and whose feedback and advice continues to make me a better researcher and writer.

Thanks to those outside Chatham House who have reviewed specific contributions to this book, including Vijay Bhopal of Scene Connect, Jessica Obeid, and a number of others who preferred to remain anonymous. Your considered feedback has immeasurably improved the content.

I am grateful to all of those who have worked with and alongside me in the Moving Energy Initiative for the last five years. Ben Good, Laura Patel and Joe Attwood at Energy 4 Impact. Mary Willcox, Mattia Vianello and Sarah Rosenberg-Jansen at Practical Action. Paul Quigley and Paul McCallion at UNHCR and many others have inspired and taught me much about this subject. I am also grateful to those in the sector who are not listed contributors in this volume but continue to work hard to improve energy solutions for those in humanitarian settings: Thomas Fohgrub and Aimee Jenks at the United Nations Institute for Training and Research (UNITAR), Kathleen Callaghy at the Global Alliance for Clean Cookstoves, Inga Buchholz at GIZ, Luc Severi at Sustainable Energy for All, and many more have continued to devote substantial portions of their own time and energy to making progress on this issue.

Thanks to all those at Routledge who have worked diligently in bringing this book through commissioning, editing, and production – in particularly to Matt Shobbrook and Annabelle Harris for their patience and for their keen attention to detail.

Most importantly, I am grateful to all of the chapter authors within this volume who have contributed their time, energy, and expertise to this collection and, by doing so, have made this a rewarding venture. I hope this collection serves – at least in part – to reward your efforts and your commitment to this important issue.

Introduction and overview

Owen Grafham

The seventh sustainable development goal commits the world to achieving 'universal access to modern energy services' for all by 2030 (SEforAll, 2017). However, not only is the world considerably off track when it comes to delivering this vision, but some sections of the world have seen a significant worsening of conditions over the last decade. Regions subject to armed conflict are the least likely to benefit from any global energy transition, while the people that have been forced to leave their homes in search of safety are some of the most vulnerable to the debilitating effects of energy poverty.

Until recently, few people working on energy access agendas in the development field paid attention to the specific issues facing displaced people or included them in their plans and projects. In the humanitarian field, energy is generally dealt with on an emergency and basic needs basis; energy services have variously taken the form of shelter, stove handouts, winterization kits, charcoal rations, solar lantern distribution and (mainly) diesel generation to power essential camp facilities. However, with the sharp rise in displacement across the world has come the realization among international actors in both fields that they must cooperate for more sustainable solutions in protracted situations of forced migration.

The first World Humanitarian Summit in May 2016 emphasised the gap in moving from relief to longer-term resilience for those in need of humanitarian assistance and the need to better support countries facing displacement crises – the majority of which face their own environmental and development challenges. In both cases, access to reliable and sustainable energy services is critical. As the chapters of this book will show, it is essential to many humanitarian priorities from keeping families warm enough to enabling livelihood opportunities, whilst investment in energy during a crisis can help a hosting country to meet its sustainable development goals (SDGs) and climate commitments.

This edited collection is the first book-length piece devoted to the subject. The authors are a mix of progressive practitioners working on displacement crises on the ground and thought-leaders in this emerging academic field of study. The book draws on a wide range of case studies, geographies and perspectives. It assesses current energy governance regimes, models of delivery and innovative solutions. By bringing together author-teams of practitioners, academics, businesses and policy makers, this collection aims to encourage interdisciplinary dialogue about the best way of approaching energy provision for displaced people and to provide a strong grounding in the subject for new students and scholars of this issue area.

The collection aims to shed light on the ways that the energy-access and humanitarian policy disciplines can be conceptualised together; on the settings in which these issues converge; and crucially, on emerging approaches and new opportunities that are being deployed to deliver energy to displaced people.

Three types of energy transition guide the narrative of this book. The first is the desired transition from current unsustainable modes of energy use to modes that are safer, cheaper and more reliable. The second is the transition of the humanitarian sector towards delivering better quality energy solutions – both for themselves and for the people they serve. The third is the low-carbon transition ambitions of the hosting countries – can investment in energy solutions for the displaced also benefit this and have other societal co-benefits?

Changing the humanitarian system is hard. Big institutions move slowly, and this is particularly the case for organisations that are under funded, under resourced and working in some of the most challenging environments in the world. But this pressure means the sector is also looking to deliver more in more cost-effective, sustainable ways. The momentum for change is growing with a critical mass of United Nations, NGO, country donor agency and private sector actors acknowledging the need to improve energy access or displaced people; and more finance than ever before targeting innovative solutions. In the development field, there is also renewed emphasis on 'leaving no one behind' – and that also applies to SDG 7.

Heat, light and power for refugees: why does it matter?

The United Nations' Refugee Agency (UNHCR) estimates that there are over 70.8 million people currently forcibly displaced – more than the populations of Australia and Canada combined and roughly equal to 1 in 108 people globally (UNHCR, 2019a).[1] Millions more are displaced by disasters, climate change, famine, or drought (IDMC, 2019). Yet displaced people have been largely invisible on the global energy access agenda and only came to the fore of discussions in the humanitarian sphere around 2018.[2]

Energy services are essential for basic human protection and dignity. Lighting, heating, cooking and powering a radio or cellphone all use forms of energy. Energy is essential for health care services, and water and sanitation provisions, and extremely useful for education and community facilities. And yet, previous estimations by this editor suggest that around 90 percent of those in camps do not have access to electricity, and around 80 percent in camps are cooking with the most basic fuel possible – normally wood (Lahn and Grafham, 2015).

This widespread deprivation has enormous health and safety impacts. It is estimated that those in situations of forced migration may be particularly vulnerable to indoor air pollution derived from cooking with the most basic fuels (Lahn and Grafham, 2015). Numerous studies have been undertaken that show how those in situations of forcible displacement often skip meals due to a lack of fuel (UNHCR, 2014a), swap food for fuel (UNHCR, 2014b) and have accidents due to lack of lighting (Lahn and Grafham, 2015). Energy is a life or death issue when people are living in temporary – often flimsy – shelter (Albadra, 2017).

Beyond the immediate health and safety benefits, energy access is also an enabler. "Thriving, and not just surviving" (UNHCR, 2019b) requires sustainable energy. Whether that means cooking, lighting a home, studying at night, keeping in touch or carrying out transactions by phone, travelling by motor vehicle, powering tools or listening to the radio – all of these activities require energy.

Having access to energy enables people to forge their own livelihoods. It might mean the difference between being able to start a business and not. For example, with power for a refrigerator, one could begin selling ice or cool drinks. For many it would increase their opportunities for self-reliance. Imagine being a barber who could plug in a razor for the first time, a miller who is able to mill with power for the first time or a tailor who is able to sew with a sewing machine for the first time. With reliable access to energy, people can exert more control over their lives.

This book aims to take a people-centred view of the potential for 'better' energy solutions – that is cooking, lighting, heating and power solutions that better meet the needs of displaced people. Contributions to this collection by Rosenberg-Jansen (Chapter 1), To and Subedi (Chapter 5), Bisaga and Hamayun (Chapter 9) and Hakspiel and Lehne (Chapter 11) take the perspective of displaced people as the starting point for examining how energy interventions can improve lives and move people up the energy access ladder.

Working towards 'green' agencies

The positive impacts of sustainable energy are also relevant for the humanitarian agencies serving forcibly displaced people. Energy allows humanitarians to power offices, clinics and water pumping facilities, and to transport equipment, food and staff. As the global caseload for humanitarian agencies continues to rise, it is striking that, at the time of writing, no major agency had a comprehensive strategy for addressing its energy use (Grafham and Lahn, 2018a). And despite the UN's stated commitment to carbon neutrality by 2020, most of the UN agencies serving displaced people are highly reliant on diesel generators (Grafham and Lahn, 2018a).

It is estimated that around 5 percent of humanitarian agency budgets goes to diesel (and related costs including for maintenance of generators, etc.) – a total which would mean that in 2017 the sector spent approximately $1.2billion on polluting fuel for off-grid electricity generation and in-country road transport alone (Grafham and Lahn, 2018a). Clear examples of best practice are emerging which offer a compelling cost and environmental case for humanitarian agencies to switch to more sustainable forms of energy. Operational savings can be created by – among other things – reducing the fuel used for on the ground transport, by changing the way that people use current energy services (i.e., air conditioning units) and by switching power generation to more renewable sources. Together these options offer a clear roadmap for the humanitarian sector to transition towards a lower-carbon future, to save money and to divert these savings to the urgent life-saving tasks that abound (Grafham and Lahn, 2018b).

And aside from their own energy consumption, agencies are one of the most important influencers on the way that energy is delivered to forcibly displaced people. When disasters and crises strike, it is often NGOs and INGOs who are the first people to respond. In this book, chapters by Mach (Chapter 2), Bellanca (Chapter 6), Vianello and Boodhna (Chapter 7) and Nixon and Gaura (Chapter 10) all proffer solutions that would improve the way the humanitarian sector deals with energy.

What is forced displacement?

This book largely concentrates on those counted in the category of people 'forcibly displaced', but the issues are relevant to a wider group in need of humanitarian assistance. The forcibly displaced category includes refugees, internally displaced people and asylum seekers and refers to those who are forced to leave their homes "as a result of persecution, conflict, violence, or human rights violations" (UNHCR, 2019a). As this book will show, this 'forcibly displaced' group often lack a range of political and economic rights and often live in considerable uncertainty about what the future will hold. Given the tumultuous situations that many of this group have fled from, it is perhaps also no surprise that many of this group have been displaced for a long time. UNHCR estimates that "78% of the 15.9 million refugees were in protracted situations at the end of 2018" (UNHCR, 2019a), whilst previous estimates from this editor suggest that the average lifespan of a refugee camp is roughly 18 years (Grafham and Lahn, 2018b). Although there is usually great political sensitivity concerning refugees' length of stay in the host country, and settlements for displaced people are rarely planned for the long-term, these statistics illustrate that delivering energy in a short-term manner is often an incredibly inefficient way to look after populations that actually end up staying for a long time (Lahn and Grafham, 2015).

Even acknowledging these high numbers, The Internal Displacement Monitoring Centre notes, for example, that "almost two-thirds more internal displacements were triggered by disasters (61 percent) than conflict and violence (39 percent) in 2018", a trend likely to continue as the impacts of climate change manifest themselves (IDMC, 2019). Many of the issues and solutions discussed in this volume could also apply to, or be adapted for, this type of crisis.

Traditional analysis would say that people migrate for new economic opportunities, or for access to education, or to reunite with family, or to escape persecution. However, in today's world it is often the case that many of these things are true at the same time. For example, for many young Syrians living in Damascus during the Syrian civil war, it has not been possible to think about the danger to life as being separate from the desire to work, to earn more money, to study or to take care of family. Years of poor governance and the dire economic situations that conflicts create are also drivers in many refugees decisions to leave their home country. In making the decision to migrate, all of these issues may be intertwined. Thus while this book follows those forcibly displaced, it recognises that such labels are reductive for the lived experience of migration (Crisp, 2017).

In Europe, and in the global development literature, there is some discussion about whether improving energy access to vulnerable populations can reduce the likelihood of migration. This question is significant for politicians and for development professionals who regularly link development spending to reductions in migration and is considered in this collection by Mach in Chapter 2 and To and Subedi in Chapter 5, who review the current evidence and highlight some of the conceptual frameworks that have defined how this relationship is generally conceived. However, it is necessary to highlight from the outset that this book focuses primarily on how energy access can be extended to those already on the move.

A global challenge

This book prioritises case studies from developing countries. Eighty-four percent of refugees are living in developing countries (UNHCR, 2019a). This percentage is even higher for IDPs. Around 16.5 million people are internally displaced across Sub-Saharan Africa, and 11 million people across the MENA region, together accounting for well over half of the global total of those internally displaced by conflict (IDMC, 2019). In Asia, Pakistan, Afghanistan and Bangladesh are among the countries hosting the highest numbers of refugees and IDPs. And while such countries shoulder the greatest burden from the current mass migrations of people, each has its own poverty and environmental fragility problems.

The political dynamics in these regions are changing – anti-refugee sentiments are part of the political landscape in – for example – Pakistan, Kenya, Tanzania and Lebanon. Some of this tension is emerging because citizens believe that refugees and migrants are a drain on the natural resources of the host country – whether this is water or energy or land (Bhardwaj and Grafham, 2018). A popular narrative is that the presence of refugees or IDPs is making an already bad situation worse, putting pressure on or causing inflation in resources such as food and energy, undercutting wages for local jobs or overstretching municipal services such as healthcare and waste management (Lahn, Grafham Sparr, 2016). In this context, host governments are often reluctant or opposed to services or infrastructure provision that implies permanence or better conditions for displaced communities than the local population.

But energy offers one means of countering this perception. In Chapter 5, To and Subedi examine an emerging framework for 'community energy resilience' – concentrating on rural communities in post-disaster Nepal. In Chapter 4, Jourdi, Gharaibeh and Qahoush look at the potential of the humanitarian sector to use energy interventions to contribute to the resilience of cities like Irbid, Jordan. Globally, around 60 percent of refugees and 80 percent of IDPs find refuge among the host population, largely in urban areas (Muggah and Abdenur, 2018). This is the situation in Jordan, with at least 80 percent of refugees living alongside the host population. Both chapters are examples of increasing attention to the 'response and resilience' agenda and how energy can play a role in such an agenda. While energy has not previously been a core priority of humanitarian relief, the importance accorded to it in Uganda, Jordan, Lebanon, Kenya and Bangladesh have made it one priority in crisis response plans. Governments identify energy as an area of stress, for which aid can help them to better cope with the new influx of people (Lahn, Grafham and Sparr, 2016).

A country's broader plan and regulations for energy access or sustainable energy transition (or lack of them) will influence what kind of energy projects are possible. Where energy is demarcated as a government priority, it opens the door to investment in energy provision as a way to secure longer-term security for refugees. In this book, contributions by Jourdi, Gharaibeh and Qahoush in Jordan (Chapter 4), Vianello and Boodhna in Burkina Faso (Chapter 7), Nixon and Gaura in Rwanda (Chapter 10) and Hakspiel and Lehne in Kenya (Chapter 11) all consider how energy interventions are situated within wider energy transition ambitions of the hosting country.

A lack of data

Studies such as those by Lahn and Grafham (2015), Rivoal and Haselip (2017) and Corbyn and Vianello (2018) have helped to shed some light on how refugees use and consume energy, but stakeholders in the sector are clear that the lack of data still amounts to one of the main barriers to improving humanitarian energy access. This is a problem because it is difficult to construct solutions if one doesn't understand the extent of the problem. Opportunities to do things differently are routinely missed because decision makers lack the requisite data on energy use, costs and alternatives. In Chapter 1 of this collection, Rosenberg-Jansen illustrates how better quality data can help us to overcome pervasive challenges and several other authors in this collection make the point that more (and higher quality) data are needed. Chapter 9 by Bisaga and Hamayun shows one way forward through a case study of the solar-home-system provider BBOXX and how this company uses large amounts of quantitative data to identify appropriate solutions.

Money and markets

Despite the lack of high-quality data in the field, we know from the small amount of quantitative work that has been done that people's desire for energy services – lighting, cool water, phone charging and power for equipment that can help people to earn a living – is strong. In most cases, refugees are already paying something for energy, and often at a high price relative to the quality of access they receive (Rivoal and Haselip, 2017). In Kakuma in Kenya, "more than one-third expressed a willingness to pay for quality household solar products, indicating a potential customer base of 5,000 families and a market worth some $300,000" (Corbyn and Vianello, 2018). The first World Humanitarian Summit in 2016 emphasised that enabling self-reliance of displaced people is a major issue. However, the ability of displaced populations to pay varies widely from place to place and from household to household and these variations need careful consideration when planning a market-based approach to energy services.

Vianello and Boodhna (Chapter 7) examine how a market-systems approach to providing energy can apply to humanitarian settings, with particular reference to work undertaken in Burkina Faso. Bellanca in Chapter 6 provides further examples of how market-based interventions can be structured, drawing on her work with Mercy Corps and the World Food Programme and looking at how energy projects can interact with increasing use of cash-based assistance. And in Chapter 11, Hakspiel and Lehne look at how more circular approaches to waste and energy can be devised that also meet user preferences and ability to pay.

Technology

A breathtaking range of innovative products and services exists to meet energy needs. But it can be argued that the humanitarian energy debate has for years been dominated by a 'technology-first' perspective (Lahn and Grafham, 2015). This has often resulted in humanitarian agencies aiming to 'fix' an energy access problem simply by providing products, rather than thinking through the user experience, maintenance arrangements and cultural habits which affect use. This collection does not seek to downplay the importance of technology, and several chapters in this collection highlight the fact that the emergence (and perhaps more importantly adaptation and refinement) of technology will be crucial to overcoming energy access barriers in humanitarian settings. Chapter 10 by Nixon and Gaura, for example, shows that technological advances such as those around pervasive computing and sensing have been used successfully in similar environments to understand energy behaviour, to support behavioural change and enable sustainable living solutions. They therefore make the case for an increased focus on such technologies, drawing on their work in Rwanda as part of their Humanitarian Engineering and Energy for Displacement project.

However, rather than examine specific technologies, this book focuses on the interplay between a given technology, its implementers and users and the governance systems that influence its success. Experience to date suggests that the most durable solutions for improving access to energy in off-grid humanitarian settings are normally 'tried and tested' in the broader rural energy access space (Grafham and Vianello, 2018). Existing projects involving mini-grids, solar-home-systems and alternative cooking fuels are highlighted in this collection. However, in all the aforementioned cases, it is not the technology that is innovative – rather, the business models, and/or new ways of structuring financial risks between philanthropic, humanitarian and private sector actors (Grafham and Vianello, 2018).

Gender

Numerous studies highlight the vulnerability of women and girls in humanitarian settings (Murphy et al., 2016; Women's Refugee Commission, 2015; Mapp, 2010), and more recent estimates suggest that around one in five refugee and displaced women in complex humanitarian emergencies have experienced sexual violence (Murphy et al., 2016; WRC, 2015).

The gender dynamics surrounding the procurement and use of energy should also be understood. Around the world, and particularly in areas of Sub-Saharan Africa where energy poverty is acute, cooking and feeding one's family remains a primary concern and the main energy-using activity for most displaced and refugee households (Lehne et al., 2016).

Women and girls are usually responsible for the collection of the firewood that fuels the majority of cooking by refugees. Leaving camps to obtain this firewood – as well as water in some cases – can expose women and girls to intimidation and even sexual violence. They also have to make the critical day-to-day decisions about whether they have enough fuel to cook their food, and who should go without if they don't. Bartering food for fuel is common practice in many camps.

Women and children are also the most vulnerable to the air pollution that can be generated by cooking on open fires (particularly those who have to cook inside their shelters). This frequently causes respiratory health and eye problems and can negatively impact on pregnancies and foetal development. The pollution caused by cooking can also reinforce inequalities and become a cause of additional stress. In camps in Kenya and Burkina Faso for example, soot is a constant problem for women and linked to status, requiring arduous cleaning (Cross et al., 2019).

And the gendered nature of energy doesn't end with cooking. Lack of lighting at night can deter women and children from going to the bathrooms or joining communal activities after dark. Whilst it is not the cause of violence, darkness can reinforce an atmosphere of insecurity. In addition, births are one of the most vulnerable times for women, these often take place at night and lighting for a midwife can be a matter of life or death. Where there are complications, or if c-sections need to be performed, power and clean water are essential.

Historically, many energy interventions in humanitarian settings have been justified on the basis of combating violence against women and girls. Cookstove projects, for example, have often been promoted on the basis of reducing the number of trips that women need to make to collect firewood – and thus avoiding harassment and gender-based violence (Global Alliance for Clean Cookstoves, 2016). Similarly, street lighting interventions have been promoted on the basis that they can make women feel safer when travelling around camps and accessing camp facilities after dark.

But this relationship is not one-dimensional. Recent studies have improve our understanding of how gender concerns can be built into humanitarian energy programming in ways which take account of the more complex context which leads to violence against women and girls. An example of this would be taking account of intimate partner violence – the most common cause of violence against women both in and beyond refugee settings – but often pervasive in camps and informal settlements with widespread unemployment, disempowerment and degraded living standards. Chapter 8 by Bradley and Liakos reviews the existing work in this space and argues for a broader and more gender-sensitive understanding of how violence against women and girls can interact with future energy projects.

Freedoms

One key areas of difference for humanitarian energy projects is the extent to which refugees are allowed to work and to move freely around the country. Most refugees have limitations imposed on their ability to work or to move freely and these restrictions have major implications for the ways in which they can engage with energy products and services.

Restrictions on leaving a camp or designated area, for example, will limit the ability of refugees and displaced people to purchase or acquire energy-access services, technologies and products that might otherwise be available. In some cases, restrictions are weakly enforced and refugees have de facto unrestricted movement, although they may be vulnerable to arrest or intimidation when they leave the camp confines. In others, there are checkpoint, monitoring and permitting systems, with the degree of enforcement varying from country to country. These systems constrain refugee movement, by setting and enforcing the conditions under which refugees are allowed to leave (Grafham and Vianello, 2018).

Whilst we know that refugees are often willing and able to travel great distances to access energy supplies (Cross et al., 2019), many of those in camps are reliant on supply chains extending long distances beyond metropolitan areas into relatively remote parts of the country. This strongly affects the distribution and availability of quality products, transport fees incurred by providers and therefore affordability and maintenance and warranty arrangements and after-sales servicing (Grafham and Vianello, 2018).

Perhaps more fundamentally, refugees often lack the legal freedom to work, especially outside of informal, barter and 'cash for work' camp economies. This can limit their ability to purchase energy products and services, and may be a barrier to successfully deploying improved electricity supply to generate income. Recent energy-access literature places great importance on 'productive-use' – i.e., how new energy sources can and are used for hitherto inaccessible productive purposes (Poor People's Energy Outlook, 2012; PRODUSE, 2011). Theoretically, lacking the freedom to work means that it should be difficult for refugees to use new supplies of energy productively, for example by setting up businesses which make use of improved electricity supply. But in practice we have found that the main barriers to refugees establishing enterprises in humanitarian settings are not legal but financial – a finding backed up by studies in other countries not specifically focused on energy (Zetter and Ruaudel, 2016). In Goudoubo refugee camp, key enterprises include rearing livestock, leatherwork, food stalls, restaurants and retail shops. The camp also has a mechanic, a tailor and two blacksmiths, but these camp entrepreneurs are limited to using hand tools (Corbyn and Vianello, 2018).

Different, but similarly vibrant markets can be found in many camp environments. Zaatari camp in Jordan is famous for it's so-called 'Champs Elysee' – a street selling everything from kebabs to wedding dresses, and where store-holders rely on diesel generators.

Whilst this book does not explicitly consider how changes in freedoms can affect the provision or take-up of energy services and products, Chapter 2 by

Eva Mach outlines how broader policy changes in the humanitarian sector (for example around freedoms and rights) make space for improved energy provision. Chapter 3 by Lahn and Grafham provides more detail on why plans to deliver energy projects to refugees and displaced people can learn much from the experience of rural or peri-urban energy access projects, and yet must be adapted in view of specific legal, rights and socio political constraints of specific settings.

The chapters

The first section of the book, 'Framing Energy Access and Forced Migration' attempts to provide an overview of the key concepts, debates and issues that have arisen in the sector. In Chapter 1, Sarah Rosenberg-Jansen of the University of Oxford provides an introduction to the overall governance of the humanitarian energy sector, including the array of institutions currently working in this space and the definitions and concepts which have permeated the field. In Chapter 2, Eva Mach of IOM shows how, at the policy level, bold commitments have also emerged from a raft of new international policy measures such as the Global Compact for Safe, Orderly and Regular Migration (GCM), the Global Compact on Refugees (GCR) and the UN's Agenda for Humanity. Mach shows how these policy developments support the energy access agenda of humanitarians and provides context for where energy sits in the broader push towards promoting refugee rights and freedoms. Chapter 3 by Glada Lahn and Owen Grafham gives an overview of the variety of energy needs which are facing forcibly displaced people around the world.

The second section of the book 'Evolving Approaches' looks at new methods that are dictating how energy is – and can be – provided to those who have been forcibly displaced. Chapter 4 by Jourdi Gharaibeh and Qahoush uses the example of urban areas in Jordan to examine how a specific set of humanitarian energy projects – namely solarising public buildings – can have an impact on improving national resilience where services such as health and education are affected by displacement. In Chapter 5, To and Subedi examine the broader concept of community energy resilience, making the case for greater attention to how this concept is defined and used. Drawing on the case study of Nepal, it illustrates how emerging learnings from the field are influencing broader conceptions of the 'resilience' narrative.

Chapter 7 by Vianello and Boodhna shows how a market-systems approach can be used in the delivery of humanitarian energy services, with certain caveats. The chapter highlights how traditional approaches to energy (pursued through short-term and ad-hoc methods of distribution) have had serious deleterious effects on refugees in Burkina Faso, and argues that new methods are possible if enough focus is granted on the wider ecosystem outside the specific displacement setting. Chapter 6 by Bellanca builds on this analysis by using her extensive practitioner experience to examine how market-based approaches have been implemented by organisations around the world. The chapter pays particular attention to examining the rise of cash-based programming and how this could be applied to energy interventions in the future.

Chapter 8 by Bradley and Liakos reviews the existing literature around how VAWG has been conceptualised in energy projects and argues for a wider and more textured understanding of how gender issues can intersect with energy, which takes into account the fact that VAWG needs to understand the behaviours and motivations of both men and women (as well as seeking to reduce opportunities where VAWG may occur). It makes a strong case for mainstreaming a VAWG-sensitive approach in the design of future humanitarian energy projects.

The final section of the book, 'Future Opportunities and Solutions', looks at three cutting-edge solutions that could transform how the humanitarian sector deals with energy-related projects. In Chapter 9, Bisaga and Hamayun look at the importance of data collection and offer a case study of the solar-home-system provider BBOXX; reflecting on how the company uses large amounts of quantitative data to identify appropriate solutions for their user-base (which is expanding to humanitarian settings). Chapter 10 by Nixon and Gaura examines the possibility of using wireless sensor networks to monitor performance of energy assets and to develop user-driven control of renewable energy micro-grid systems. And Chapter 11 by Hakspiel and Lehne examines how the notion of the circular economy can be applied to refugee camps, drawing on the specific example of the waste-to-energy company Sanivation, who work in the Kakuma refugee camp in Kenya.

Notes

1 These included 25.9 million refugees: 20.4 million under UNHCR's mandate and 5.5 million Palestinian refugees registered with the United Nations Relief and Works Agency for Palestine Refugees in the Near East (UNRWA). The global figure also included 41.3 million internally displaced persons (source: IDMC) and 3.5 million individuals whose asylum applications had not yet been adjudicated by the end of the reporting period. See UNHCR (2019b), Global Trends: Forced Displacement in 2018, *Office of the United Nations High Commissioner for Refugees*

2 Displaced people and refugees featured for the first time in global discussions of the United Nations Sustainable Energy Access for All (SEforAll) agenda – first at a side meeting in New York in 2017, then as a mainstage event in 2018. At the same time, new policy processes such as the Global Plan of Action for Sustainable Energy Solutions in Situations of Displacement (GPA), launched in July 2018, have begun to mainstream the issue within the UN family and within other humanitarian agencies.

Bibliography

Albadra, D. et al. (2017), 'Thermal Comfort in Desert Refugee Camps: An Interdisciplinary Approach', *Building and Environment*, 124, 460–477.

Bhardwaj, G. and Grafham, O. (2018), 'Why access to energy can empower refugees', *Chatham House*, London, UK, https://www.chathamhouse.org/expert/comment/why-access-energy-can-empower-refugees.

Corbyn, D. and Vianello, M. (2018), 'Prices, Products and Priorities: Meeting Refugees Energy Needs in Burkina Faso and Kenya', *Chatham House*, London.

Crisp, J. (2017), 'International Migration, Refugees, and Human Displacement: Some Frequently Asked Questions', Presentation to the Queen Elizabeth II Academy, *Chatham House*, 14 June.

Cross, J. et al. (2019), 'Energy and Displacement in Eight Objects: Insights from Stories in Sub-Saharan Africa', *Chatham House*, London.

EUEI PDF (2011), 'Productive Use of Energy (PRODUSE): A Manual for Electrification Practitioners', *European Union Energy Initiative Partnership Dialogue Facility (EUEI PDF)*.

Global Alliance for Clean Cookstoves (2016), 'Gender-Based Violence in Humanitarian Settings: Cookstoves and Fuels'.

Grafham, O. and Lahn, G. (2018a), 'The Costs of Fuelling Humanitarian Aid', *Chatham House*, London.

Grafham, O. and Lahn, G. (2018b), 'Powering Ahead: Improving how We Use and Account for Energy in Humanitarian Operations', *Chatham House*, London.

Grafham, O. and Vianello, M. (2018), 'Energy for Displacement: Key Issues', Cited in 'Energy for Displacement "Understanding the Challenges"', HEED Briefing Paper No.1, Humanitarian Engineering and Energy for Displacement Project, *Scene Connect.*

IDMC (2019), '2019 Global Report on Internal Displacement', *Internal Displacement Monitoring Centre*, Geneva, Switzerland.

Lahn, G. and Grafham, O. (2015), 'Heat, Light and Power for Refugees: Saving Lives, Reducing Costs', *Chatham House*, London.

Lahn, G., Grafham, O. and Sparr, A. (2016), 'Refugees and Energy Resilience in Jordan', *Chatham House*, London.

Lehne, J. et al. (2016) 'Energy services for refugees and displaced people', *Energy Strategy Review*, 13–14, 134–146.

Long, K. (2013), 'The Point of No Return: Refugees, Rights, and Repatriation', *Oxford University Press*.

Mapp, S. C. (2010), 'Global Child Welfare and Well-Being', *Oxford University Press*.

Muggah, R. and Abdenur, A. E. (2018), 'Refugees and the City the Twenty-First-Century Front Line', World Refugee Council Research Paper No. 2 – July 2018, *Centre for International Governance Innovation.*

Murphy, M. et al. (2016), 'Evidence Brief: What Works to Prevent and Respond to Violence Against Women and Girls in Conflict and Humanitarian Settings?', *The Global Women's Institute and International Rescue Committee*, George Washington University and London.

Practical Action (2012), 'Poor People's Energy Outlook, Energy for Earning a Living', *Practical Action Publishing*, Rugby, UK.

Rivoal, M. and Haselip, J. (2017), 'The True Cost of Using Traditional Fuels in a Humanitarian Setting: Case Study of the Nyarugusu Refugee Camp, Kigoma Region, Tanzania', UNEP DTU Working Paper Series 2017: 3. *UNEP DTU Partnership*, Technical University of Denmark.

SEforAll (2017), 'Going Further, Faster, Together', 2017 Annual Report, *Sustainable Energy for All*, Vienna, Austria.

UNHCR (2014a) 'Light Years Ahead Project: Monitoring and Evaluation System and Baseline Assessment Report – Chad Country Report', July 2014, *Office of the United Nations High Commissioner for Refugees*, Geneva, Switzerland.

UNHCR (2014b), 'Light Years Ahead Project: Monitoring and Evaluation System and Baseline Assessment Report – Uganda Country Report', July 2014, *Office of the United Nations High Commissioner for Refugees*, Geneva, Switzerland.

UNHCR (2018), 'Global Trends: Forced Displacement in 2017', *Office of the United Nations High Commissioner for Refugees*, Geneva, Switzerland.

UNHCR (2019a), 'Global Trends: Forced Displacement in 2018', *Office of the United Nations High Commissioner for Refugees*, Geneva, Switzerland.

UNHCR (2019b), 'A Message to the World's Refugees from UNHCR Chief Filippo Grandi', 20 June 2019, *Office of the United Nations High Commissioner for Refugees*, Geneva, Switzerland.

WRC (2015), 'Protecting Women and the Environment in the Great Lakes Region', *Women's Refugee Commission*, New York.

Zetter, R. and Ruaudel, H. (2016), 'Refugees' Right to Work and Access to Labor Markets – An Assessment', *Global Knowledge Partnership on Migration and Development (KNOMAD).*

Part 1

Framing energy-access and forced migration

1 Leaving no one behind

An overview of governance of the humanitarian energy sector

Sarah Rosenberg-Jansen

Introduction

Understanding humanitarian energy issues requires consideration of the structures of governance, institutions and practices involved in energy for displacement. This chapter will provide a brief overview of the emergence of the humanitarian energy sector, including how institutions such as UNHCR, global bodies, funders and implementing partners function, and the international and local levels at which governance and ideas are being shared. The aim of chapter is provide an overview of the governance and policies within humanitarian energy, while opening up discussion on key challenges, barriers and solutions.

The issue of energy in displaced settings is relatively new, with the first substantive research, policy and programmatic action in this area only starting in 2014 (Gunning, 2014 and Lehne et al., 2016). Prior to this, there was a lack of detailed consideration of energy needs within displaced settings, with most previous research focusing on cooking practices in situations of displacement and very few substantial on-the-ground programmes.

Humanitarian energy can be seen as a point of intersection between two sectors: decentralised energy for development and humanitarian response. The energy access sector works over the long-term in developing countries – often focusing on decentralised sustainable energy and off-grid solutions and working at the national and local levels (Practical Action, 2018 and SEforALL, 2018). The humanitarian response sector, however, is globally led by UN bodies, large NGOs, and emergency responders to focus on short-term crisis such as conflicts and natural disasters. Neither of these sectors has fully embraced or embedded energy for displaced people, and as a result there are often frictions (or a stark lack of communication) between emergency response institutions and longer-term development agencies in emerging projects and policies.

Several policy initiatives are now trying to bring the learning and systems of these two worlds together to effectively provide energy within humanitarian contexts (Lahn and Grafham, 2015; Grafham et al., 2016), but different governance regimes and varying institutional policies between the two sectors make this challenging. The sheer range of donors, implementing agencies, NGOs, national and local governments, as well as the needs of populations of concern, make policy and practice within humanitarian energy complex.

Understanding how policies and practices for humanitarian energy are supporting global objectives to deliver sustainable energy for all is vital to move the sector forward. For example, embedding new humanitarian energy strategies within the context of energy access sector learning on greening mini-grid supply, the social impacts of productive uses, and understanding market needs in national contexts (Colenbrander et al., 2018; SEforALL, 2017). This chapter presents an overview of the institutions, concepts, and organisations acting within the sector to lay the ground for detailed discussion in the following chapters.

What is energy and how is it different to other sectors?

The essential need for energy in humanitarian contexts

There is an urgent need to reshape the way energy resources are dealt with in humanitarian contexts as multiple, complex, and repeated humanitarian crises are becoming more frequent and there is a critical need for multi-year humanitarian planning to address immediate and long-term energy needs of displaced people. Energy services are key to enabling displaced people to escape poverty and delivering objectives on resilience, climate change, long-term responses to protracted crisis, market-based durable solutions, supporting livelihoods and productive opportunities, and improving the lives of women and girls.

Without energy, families cannot cook their food, bring lighting into their homes, heat and light their homes, or power their businesses. Communities cannot power their health centres, schools, or public spaces. Energy enables, but is also essential for displacement settings. For example, in humanitarian contexts electricity is essential for powering reception centres during emergency response and is needed for homes, schools, and businesses to enable people to live dignified and productive lives. In addition, energy which comes from renewable or sustainable source can reduce the environmental and resource burden on host governments and local communities and provide affordable electricity for entrepreneurs to develop livelihood opportunities based on sustainable energy. Without access to sustainable energy the options for building programmes focused on self-reliance of communities and improving livelihoods are limited.

While it is challenging to integrate sustainable energy at the start of humanitarian programming, it is possible. Energy planning tools, technical guidelines, business models, and procedures already exist to support humanitarian assistance in delivering sustainable energy. Working in partnership across humanitarian and development organisations is critical to delivering this.

Definitions: what is energy for displacement?

This chapter presents an overview of how policies and practices are developing to support cost-effective, environmentally sustainable energy to meet the needs of displaced people. Many organizations are now working on a common vision to ensure that "Every person affected by conflict or natural disaster has access to affordable, reliable, sustainable and modern energy services by 2030" (Global Plan of Action for Sustainable Energy Solutions in Situations of Displacement, 2018, p. 2). However, few of the organisations working on humanitarian energy define *what* exactly energy for displacement means. As the humanitarian energy sector expands, terminology is likely to become increasingly important. There are a range of terms employed within the sector, of which 'energy for displacement', 'safe access to fuel and energy in humanitarian settings', and 'humanitarian energy' are commonly used – often interchangeably by policy makers and practitioners.

While there is no agreed upon definition of energy for displacement, this chapter suggests the following definition for the humanitarian energy sector: *Institutions, policies, programmes, global initiatives, actions, and activities which use a range of sustainable and fossil fuel energy sources in contexts of displacement, to meet the energy needs of people in camps and urban settings, self-settled refugees, host communities, and internally displaced people.* This definition may be used as an umbrella term for a range of topics, including energy for displacement, energy for refugees, energy in emergencies, and humanitarian energy, to fully capture the dynamic scope of the terminology needed.

Table 1.1 presents an overview of the commonly used terms in the humanitarian sector to describe energy activities. While this table is not meant to be comprehensive, it does provide a sense of the varied elements of the humanitarian energy system and the core areas of focus to date.

The reasons for the differences in terminology and use of these definitions are complex, however, many of these variations are guided by two factors 1) the scope and ethos of the institutions who have been key in developing the policies and outputs for the area and 2) the emerging and changing nature of the sector. Many actors are still engaged with justifying the need and purpose of energy for displacement, and as a result the terms used to describe the problem are often adjusted to fit the audience. For example, the focus on safe, healthy, affordable fuels for cooking in humanitarian settings may be guided by the Global Alliance for Clean Cooking's remit to "save lives, improve livelihoods, empower women, and protect the environment by creating a thriving global market for clean and efficient household cooking solutions" (Global Alliance for Clean Cookstoves, 2014). Similarly, certain terms focus on technology-specific areas, such as cooking systems in humanitarian settings or the replacement of diesel generators. Even from this brief overview it is apparent that there are multiple terms and practices present within the humanitarian energy nexus.

How is energy different than other sectors?

Until recently, energy has largely been overlooked as an issue within humanitarian communities. On the one hand, energy provision in humanitarian settings is not new: there has always been a need for diesel fuel and generators to power humanitarian operations, fuel for trucks, cars, and planes to field and rural operations, and energy supplies needed to power humanitarian responses. In many humanitarian

Table 1.1 Suggested definitions and scope of humanitarian energy areas

Commonly Used Terms	*Current scope*	*Suggested Definition*
Energy for displacement, energy for displaced people, energy and forced displacement	Sector-wide term, holistic and covering a range of displacement contexts. Energy for displacement is often the umbrella and informal term used across settings.	Energy supply for displaced people: forced migrants, including refugees and internally displaced people, in camps and urban settings. Host communities and self-settled refugees can also be included in this definition as their energy needs are impacted by displacement, although these terms normally focus on energy for people in refugee camps or urban areas. These terms are often used interchangeably within the sector.
Humanitarian energy: energy access in humanitarian settings	Sector-wide term, holistic and covering a range of humanitarian contexts. Increasingly, practitioners are using the term humanitarian energy instead of energy for displacement to align explicitly with the humanitarian sector.	The use of a range of energy sources across all contexts of displacement, which includes the energy needs of people in camps and urban settings, self-settled refugees, host communities and internally displaced people. While energy for displacement is often used as a neutral term, humanitarian energy is often being used to align with progressive ideals on renewable energy and emerging lessons from energy access sector on energy access rights, sustainability, and leaving no-one behind in the transition to modern energy access.
Energy in emergencies	Focused on crisis and emergency situations, including natural disasters and conflict settings.	The use and supply of energy for crisis situations, including natural disasters and conflict settings. Often focused on short-term power and fuel supplies and the immediate needs of newly encamped populations. Energy in emergencies can also cover a wide range of people, included those who are not displaced from their region, but are in an emergency in or near their home (for example, people affected by earthquakes or natural disasters).
Energy and migration	Focused on people on the move and migrants.	The supply for, and use of, energy by migrant communities, including economic migrants and asylum seekers. Often focused on migrants to the global north rather than migrants within the global south or internally displaced people, and linked with climate change and development narratives (see following chapter by Eva Mach).

Renewable energy for refugees	Renewable energy focus and usually supporting refugees and increasingly host communities.	The use of renewable, sustainable and replenishable biomass energy sources in contexts of displacement. This terms often is used in association with refugee camps, as camps are often densely populated and renewable solutions can be implemented within communities.
Cooking and cooking systems in humanitarian settings	Cooking, firewood, and fuel focused.	The use of firewood and fuels as energy for refugees and displaced people: the supply for cookstoves, clean cooking technologies and firewood for displaced people.
Fuel and energy, safe access to fuel and energy in humanitarian settings	Focus on fuel and energy: firewood for households and diesel generation for humanitarian response organisations.	Sustainable and safe use of firewood and fuels as energy for refugees and displaced people and effective use of energy resources by humanitarian agencies.
Household energy for refugees	Focus on lighting, mobile charging, heating, cooking, and cooling for homes and households.	Ensuring refugees in camps and urban settings have safe and sustainable energy for their household use, including affordable lighting and mobile charging, safe cooking technologies, low-carbon power for household appliances.
Community energy for refugees	Often focused on mini-grid and decentralised energy solutions.	Providing sustainable and affordable energy to community facilities, including schools, hospitals, community spaces, and street lighting. Often focuses on electricity solutions, but can encompass community cooking initiatives and kitchens.
Energy for productive use in humanitarian settings and energy for refugee businesses	Usually focused on camps, mini-grid solutions, and electricity-based solutions. An emerging set of terms that may increase in visibility over time.	The supply of energy for small and medium sized enterprises in refugee camps: enabling entrepreneurship to develop within refugee camps. Often involving commercial, agricultural, and industrial activities using electricity services for the production of goods or provision of services.
Institutional energy for humanitarian agencies, energy for operations	Focused on camps, grid-level connection services or and mini-grid solutions.	The supply of energy for humanitarian agencies, NGOs and implementing partners working in refugee camps and crisis response situations. Focused on electricity products and services: currently most operations use diesel fuel to power their operations.

Source: Author's own.

responses, cookstoves, firewood and basic provisions are supplied to displaced households for cooking food and boiling water. Some humanitarian locations are even connected to grid-level power from national sources or use electricity from host communities. However, supplying renewable or sustainable energy solutions for refugees and humanitarian operations is in its infancy, as is the provision of electricity services and products in responses to enable households to have access to light and sometimes phone charging. In this sense, renewable and sustainable energy services remain exceptions rather than the rule in humanitarian contexts.

Humanitarian energy is both new as an area of policy and practice and different to other existing humanitarian sectors. There are many complex reasons for this; however, one critical reason is likely to be that there is not an energy cluster within the UN OCHA cluster system (2018). Supported by the Inter-Agency Standing Committee (IASC), the UN clusters are groups of humanitarian organizations coordinating across the core humanitarian sectors, e.g., nutrition and food, water, protection, health, early recovery, and logistics. One organisation is selected as the cluster lead and works to coordinate clear responsibilities during humanitarian emergencies for response programming (Humanitarian Response, 2018). Currently energy is not listed as a cluster, however, many core services such as camp coordination and management, planning and logistics, telecommunications, and sanitation services require energy. The absence of an energy cluster energy suggests that institutions and operations in the humanitarian energy sector have not had a clear forum for coordination and funding, nor have they had a mechanism for sharing learning on how to provide energy in displacement settings. The lack of a cluster for energy is cited by professionals in the sector as an important reason why energy needs are only now being documented as a significant topic in humanitarian contexts.

However, it could be that other reasons may be the *cause* of energy not being considered a core element of humanitarian response, and that not having an energy cluster is more of a symptom. One of the core reasons cited by humanitarians for the lack of inclusion of energy as part of core agency programming is that responses are often framed as emergencies: when a catastrophe or war strikes, many humanitarians suggest that there is no time to contemplate or design solutions for energy needs. This argument is also made concerning funding needs within humanitarian response: some institutions argue that there is not enough money available to consider energy needs of households beyond firewood and diesel supplies for response coordinators, especially at the moment of emergency. At the heart of this view is the opinion is that there isn't money and there isn't time for energy to be a priority.

Energy is different to other humanitarian sectors in this way because it is not seen as a basic right or an important resource that displaced people can draw upon or need, as food or water might be. The core argument for providing shelter, protection, WASH, and food supplies in humanitarian settings is that displaced people need these things to survive. This implies that the complexities of humanitarian operations are perhaps obscuring the lived realities of displaced people: food is important but people can't eat it without it being cooked; water is essential to life but it normally needs energy to be pumped or heated, and education is vital, but many of the basic services needed for education, livelihoods, and quality of life can't be provided without electricity. Energy is different to other humanitarian issues because the clear link between provision of the goods and why the goods are needed is often obscured. Recent policy initiatives have made the case that energy is also essential for refugees and displaced people and that energy is a human right (Lahn and Grafham, 2015). The recognition of this has led to increasing numbers of sustainable and renewable energy projects both for humanitarian operations and for displaced people themselves.

While there are of course important and practical reasons why energy is not aligned with the other humanitarian sectors, the fact that energy is a highly technical and specialised sector is also a factor. While building latrines, supplying food and hygiene kits can often be solved using logistical skills, energy supply, and technologies such as generators, invertors and grid power often require complex technical skills to connect and can be dangerous for nonspecialists. Energy is not a core competency for the humanitarian sector, and there are relatively few NGOs or implementing partners who traditionally work with UNHCR (the UN agency with a core remit for refugees) and response agencies who are specialists in energy. As a result, it can be difficult for the humanitarian system to provide and procure sustainable energy through its usual routes. Many humanitarian organisations outsource technical skills to implementing partner organisations and this could also be one reason why there is little 'in-house' specialism on energy in key agencies.

A further constraint to action is that refugee populations are often considered temporary residents of national governments. Many humanitarian organisations suggest that providing electricity services can look like long-term infrastructure investments in displaced settings. In the past this has been seen as a negative, as these services can be viewed as permanent connections by host communities which visually demonstrate the support that displaced people are getting. In some locations, the issue of whether refugees have more resources and access to services than the local population has been identified as a source of friction.

Some humanitarian institutions have been limited in their approach to energy, as the responsibility for energy supply is understood to be the remit of national governments. Recently, there appears to have been a change in perceptions in humanitarian institutions and national bodies on this topic, with many countries such as Uganda and Rwanda using an alternative approach to hosting refugees by integrating refugee responses within host community settings and enabling energy services as part of this transition. Indeed, many hosting governments seem to recognise that improving the energy supply to humanitarian settings also provides opportunities for extending this supply to local populations.

Elements of this change may be caused by the recognition that most displaced people have been away from their country for so long that the possibility of people 'returning home' is drastically reduced. There are several examples of very long-term, protracted situations of displacement. For example, Palestinian refugees who have been in Irbid camp and Zarqa camps in Jordan since 1947, people in Kenya who have been in Kakuma camp for 26 years and in Dadaab since 1991, the Rohingya people in Bangladesh who have been stateless since 1982, and Congolese refugees who have been in Rwanda since 1996 (Rosenberg-Jansen, 2018a). It is suggested that the average age of a refugee camp is over 18 years (Grafham and Lahn, 2018). Even in more recent emergencies, such as the war in Syria, which has now been running longer than seven years, the possibilities for returning to a region with almost no remaining basic utilities and energy services appear to be slim.

In short, there is an increasing recognition that modern energy is an essential requirement and a basic need for these populations to support long-term integrated development which benefits both displaced people and host communities (Grandi, 2017). Policy papers and international initiatives such as the Comprehensive Refugee Response Framework and the Global Compact on Refugees (UNHCR, 2017) now support longer-term and integrated methods for livelihood opportunities, as well as promoting co-benefits of health and education services for host communities. Renewable and clean energy supply is progressively viewed as one of the critical enablers essential in delivering change for displaced people (UNITAR, 2018); the policies and programmes underpinning this change are explored by Eva Mach in the following chapter. The next section turns to the global policies and practices within humanitarian energy and which organisations and institutions are supporting energy for displacement.

Policy and practice: progress at the international level

International action and agency policies

While the purpose of this chapter is not to provide a detailed history of humanitarian energy, and the following chapter will talk about the evolution of the international policy framework in more detail, this section will provide a short overview of how the policies of humanitarian agencies have shaped the sector over the past decade. To date the scope of the humanitarian energy sector has largely been determined by current global policy discussions and practitioner projects. The continuing swirl of policy discussions continues to reframe humanitarian energy issues with far reaching implications.

To some extent, policy makers have been striving to improve cooking access for refugees and reducing the environmental impacts of refugee camps since the 1990s (ITDG, 1996; UNHCR, 1996). However, support for electricity access and wider considerations on humanitarian energy have been slow to take off. Dedicated policy research by Chatham House was undertaken in 2013 on the energy needs of populations in camps (Gunning, 2014) and the policies and practices of sustainable energy provision (Bellanca, 2014). Prior to this, published work on humanitarian energy largely focused on evaluations of clean cooking interventions in specific sites, for example work done on cooking and fuels in Uganda (Birikadde et al., 2009; Bizzarri and Barbelet, 2011), Kenya (Bates et al., 2002; Bizzarri et al., 2010), and Ethiopia (Egziabher et al., 2006; Rogers et al., 2013).

Since 2013 the number of projects, policies and international events, and discussions have increased rapidly.

One of the reasons for the growth of the sector could be the increased attention to energy within the international development agenda. For example, both the inclusion of energy as a Sustainable Development Goal (Nakicenovic et al., 2013; UNDESA, 2018) and the evolution of the energy access agenda (SEforALL, 2018) have increased attention on energy both as a resource and as a human right. Additionally, initial research and investment projects have been funded both by the UK Department for International Development and the IKEA Foundation (2017) to produce pilot projects that demonstrate new models for sustainable energy investments in humanitarian settings.

Within the UN and humanitarian systems, however, the energy agenda has been slow to take off. Several policy initiatives, for example the Moving Energy Initiative (2017) and the SAFE initiative (2015) have managed to create some traction with the large humanitarian agencies, and UNHCR launched a strategy for energy in 2014 (UNHCR, 2014) and have recently launched a new progressive global energy strategy in 2019 (UNHCR 2019). However, energy is still not formally included in the humanitarian response system: as discussed earlier, energy does not yet have an institutional presence within the Inter-Agency Standing Committee clusters leads system.

Figure 1.1 highlights a number of institutional and practitioner policies, frameworks and tools that have been used to address some aspects of humanitarian energy. While these policies have not been comprehensive for the policy makers or practitioners involved, they have provided some structure and guidance on key issues within agency priorities. Some of these tools are principles-based and provide general guidance, such as the UNHCR Light Years Ahead and environmental guidelines, while others are designed to support programmatic action. For example, the SAFE handbook provides decision-tree diagrams which aim to provide more concrete guidance for practitioners. As Bellanca puts it,

> the creation of the Safe Access to Fuel and Energy (SAFE) initiative has signalled a turning point by catalysing the participation of many UN agencies and NGOs in energy issues. While the SAFE coalition chiefly seeks to address energy access, other trends have been emerging in recent years that focus on how best to 'green' procurement and operations in the UN.
>
> (Bellanca, 2014, p. 2)

Alongside these policies and guidelines, there are an increasing number of pilots and policy research on humanitarian energy which are now producing initial results, data, and information. As a result, the SAFE initiative has launched an Energy Community of Practice which provides summaries and information on many of the projects emerging at the field and country levels (SAFE Humanitarian Working Group, 2017). During 2018, a Global Plan of Action on humanitarian energy was developed by a range of global institutions to move this agenda

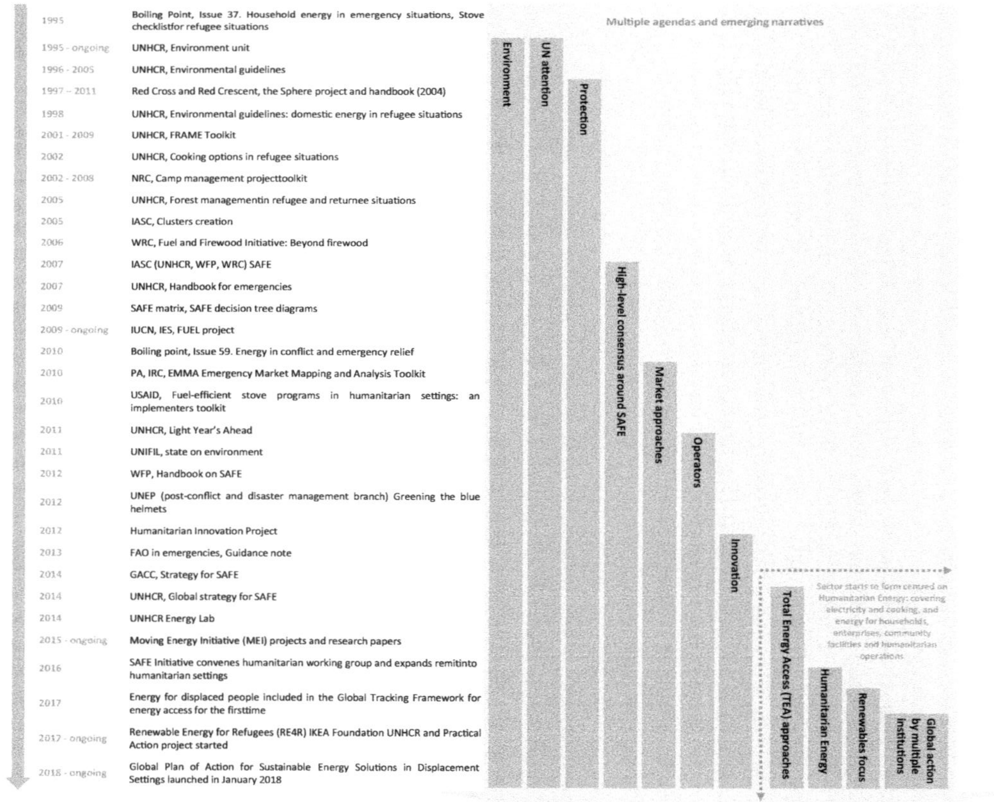

Figure 1.1 Chronological summary of events relevant to energy in emergency situations, 1995 to 2018

Source: Adapted from Bellanca, 2014, p. 6.

forward (GPA, 2018). These initiatives indicate the global level of attention this issue is now receiving and the increasing number of participants involved in the debate. Initially these policies were guided by environmental, UN systematic changes, and protection motivations. However, in recent years, the discussion has shifted towards sustainable and safe access to fuels, market approaches, innovation, and the need for humanitarian operations to reduce their costs and use of resources. Progress for the humanitarian energy agenda is now increasingly tied to two joint political agendas – the universal energy access by 2030 objective and the many drivers around efficiency and 'value-for-money' pushing the humanitarian aid sector towards internal reform.

Summary of progress: embedding energy access and humanitarian aid reform

To some degree, there has been progress to draw attention to humanitarian energy issue in international; political fora. In particular, progress has been made in the energy access sector – with humanitarian energy now often discussed alongside the Sustainable Development Goal 7 objectives "to ensure access to affordable, reliable, sustainable and modern energy for all, increase the share of renewable energy and energy efficiency" (GPA, 2018, p. 7). To date, the Global Plan of Action for Sustainable Energy Solutions in Situations of Displacement (2018) has been advocating to integrate energy needs within both international agendas, governmental practices, and institutions practices, suggesting that: "when it comes to the supply of clean and sustainable energy, governments are able to include displaced people within their national development plans and NDCs, thus bundling projects into multilateral climate change funding mechanisms" (GPA, 2018, p. 7). In the energy access world at least, it seems as if "people are beginning to realise that this is an important issue, and something that deserves priority, resources and attention" (Grafham, cited in Rowling, 2018).

Similarly, the inclusion of data on displaced people in the Global Tracking Framework (GTF) 2017 report (World Bank, 2017) represents a considerable step in highlighting the importance of evidence needs in this area, but it is just a start, and considerably more detailed evidence is critical for decision-makers to be able to deliver sustainable energy as part of humanitarian programmes and projects. For example, targets and data on energy for displaced people should be embedded within the GTF targets and tracking mechanisms to ensure that regular data is available and that progress for displaced people is monitored alongside national rural and urban communities. Some UNHCR and SAFE policies, such as those outlined earlier, have been discussed within agencies, energy is still not high on the agenda for many humanitarians: "Electric power has not been regarded as a human 'right' in emergency situations, unlike shelter, water, food or healthcare. . . . Energy is really not something that is fully taken into account, but it is the key to empowering refugees and displaced persons" (Andrew Harper – UNHCR's director of programme support, cited in Rowling, 2018).

While progress is clearly happening at the global and political levels, on-the-ground action and institutional change is required to ensure real progress is made and that sustainable energy policies and practices are embedded within organisations. As emphasised earlier, there are many structural challenges for humanitarian agencies to embedding energy into their programmes and policies in a concrete way, including technical capacity, limited funding, and agencies having to work through systems and processes which do not consider energy or support the transition to sustainable fuel sources. Humanitarian agencies and their donors may need to reform their systems and policies in order to truly deliver energy access objectives – while this may be challenging in the current political and economic climates, there is considerable support and expertise on energy available outside humanitarian institutions, and working through the types of partnerships outlined next may offer agencies support and skills in improving on delivering their mandates. Further work is still needed by intuitions and individuals to ensure humanitarian energy solutions are delivered and are evidence-based (Rosenberg-Jansen, 2018b). It is to progress in these areas that this chapter now turns.

Institutions in humanitarian energy and how they act

Institutions and organisations: structures and roles

Energy governance can take place at multiple levels, including national and international action to manage energy resources and provide energy services (Florini and Sovacool, 2009). Humanitarian resilience can also be fostered through a range of governance mechanisms, national government action and community-based initiatives (Hilhorst et al., 2019). In a humanitarian energy context, both collective and disparate action on energy is taken by a range of actors and institutions, including humanitarian response agencies, UN bodies, local and global NGOs, and think tanks. Institutional action on humanitarian energy covers projects and programmes implemented in countries and at local sites within camps, urban areas, and self-settled regions.

The institutions and modes of organisation within refugee camps are often quite different to those supporting internally displaced populations. For example, UNHCR and organisations that manage camp operations are usually responsible for overall coordination for camp activities which includes shelter, registration, provision of resources, and resettlement processes. In camps they often partner with the World Food Programme (WFP) who are responsible for food distribution, and NGOs such as World Vision, the American Refugee Committee, or the Norwegian Refugee Council who lead on delivering health, education, and community services. In terms of energy projects, UNHCR often look to work with partners, NGOs, and other investors. This is often done through implementing partners (IPs), who are local organisations and often NGOs or charities who undertake the direct work with communities in the camps and who are mainly responsible for on-the-ground delivery. In some cases, financial investment for energy projects comes into UNHCR or the camp operator and they sub-contract implementing partners. In others, money flows directly into the partner agency, and they work with the camp operators on permissions and structuring the programmes.

However, the system supporting urban refugees, self-settled communities, and internally displaced people has additional complex factors: with considerable pressure falling on national governments, local and national NGOs, and host communities. In many countries in the Middle East for example, humanitarian agencies often have a limited role in cities and towns hosting refugees. Refugees and displaced people find private accommodation, pay rent and utility bills, and use state hospitals, schools, and institutions to cover basic needs for education, health, and access to services. In the case of non-camp and non-refugee populations, the role of nation states is critical. As mentioned previously, some governments, such as those in Rwanda and Uganda, are taking an integrated host and refugee approach to managing displaced populations, which is supporting the move to long-term settlement and integration within the host communities. In these cases, responsibility for sustainable energy is increasingly being taken on by national ministries for industry, energy, and infrastructure, with support from NGOs and international development organisations. How to provide energy resources becomes even more complex when we consider internally displaced people who are displaced either by conflict or natural disaster – as often global humanitarian organisations fill the role of nation states during such crises.

Clearly, no one actor or set of actors can be said to govern the field of humanitarian energy in the traditional sense of control or mandated regulation. However, several key institutions have been involved in the shaping of the energy agenda in humanitarian settings. The following summary is not designed to be comprehensive but rather provide an overview of the key organisations who have already been involved with the humanitarian energy sector so far:

- Intergovernmental and UN bodies, such as the Food and Agriculture Organisation (FAO), International Organisation on Migration (IOM), United Nations High Commissioner for Refugees (UNHCR), United Nations Institute for Training and Research (UNITAR), United Nations Development Programme (UNDP), United Nations Environment Programme (UNEP), United Nations International Children's Emergency Fund (UNICEF), United Nations Industrial Development Organisation (UNIDO), World Food Programme (WFP), and International Renewable Energy Agency (IRENA). While a few of these organisations, UNHCR and IOM notably, have energy programmes, these are extremely small compared to the core remit of these of the organisations. Increasingly, UN bodies have been engaging with humanitarian energy at the global level, but on-the-ground programming is still nascent. The role of these global institutions is largely to provide access and coordination for energy services, rather than leading or supplying energy services.
- International and local NGOs such as Mercy Corps, MSF, NRC, Plan International, Practical Action, Oxfam, World Vision. These organisations tend to

be viewed as implementing partners who work with humanitarian agencies to support relief, response and often long-term development objectives – and are often viewed as the implementers for energy access programming. Many of these organisations already have energy programmes embedded within their existing programming, such as solar water pumping projects, but are usually more focused on other sectors such as health, food relief, or water and sanitation.

- Nation states and local governments. For example, the governments of Jordan, Rwanda, Kenya, and Uganda have committed to supporting refugees within the host communities for long-term settlements. Similarly, there are some institutions to support the energy needs of IDPs within the national bodies of Yemen, Sudan, South Sudan, Somalia, DRC, Nigeria, Colombia, Turkey, Afghanistan, Iraq, and Syria – all of which have more than a million people who are internally displaced (IDMC, 2018). The specific roles of national ministries and governments is complex in terms of energy supply for displaced people and involve a range of line ministries and implementing agencies and services delivery agents. In extreme cases, such as South Sudan and Yemen, pressures on the national government mean that energy services – alongside many other core national services – are not available for the majority of people living in these countries.
- Think tanks, funded partnerships, and policy bodies: such as Safe Access to Fuel and Energy initiative (SAFE), the Global Alliance for Clean Cooking (GACC), Energy for Development partnership (ENDEV), Alliance for Rural Electrification (ARE), Global Off-Grid Lighting Association (GOGLA), Sustainable Energy for All initiative (SEforALL), World LPG Association (WLPGA), and the Moving Energy Initiative (MEI). Many of these organisations currently have dedicated programmes or individuals working on energy for displacement in global and international forums, and have been influential in the development of the humanitarian sector. For example – the first substantive research was undertaken by MEI and Chatham House, and the SAFE working group on humanitarian energy access was the first group to convene a large number of implementation and policy actors to progress the global humanitarian energy agenda. Many of these organisations are usually focused on long-term development or energy sector issues rather than emergency response.
- Funders such as the large national donor agencies Deutsche Gesellschaft für Internationale Zusammenarbeit (GIZ), Stichting Nederlandse Vrijwilligers (SNV), the Swedish International Development Cooperation Agency (SIDA), and the UK Department for International Development (DFID); international finance institutions like the Asian Development Bank (ADB), African Development Bank (AfDB), Islamic Development Bank, KfW Bank Group, and the World Bank; and multi-donor and philanthropy organisations such as the European Union, Economic Community of West African States (Ecowas), the Gates Foundation, IKEA Foundation, Kofi Annan Foundation, and the Olympic Committee. The majority of these donors have already

funded projects in the humanitarian energy sector, with the others considering investment programmes.

- Businesses and investors from the private sector, such as Eurelectric, Energias de Portugal (EDP), Schneider Electric, Phillips Lighting, Total, Caterpillar, Inyenyeri, Little Sun, Kube Energy, BBOX, Mobisol, and PWC. The role of these organisations is highly varied: some are energy suppliers and have business in refugee camps such as Mobisol, BBOXX and Inyenyeri, while others are large energy sector companies considering market options and the potential to expand into humanitarian markets.
- Finally, there are a few academic and research organisations engaged in the sector, for example: the Massachusetts Institute for Technology (MIT), Coventry University, Engineers Without Borders (EWB), Climate-KIC, University of Oxford, and Loughborough University.

Responsibilities, forms of governance, and partnerships

Governance, regulation and institutional engagements would normally include oversight of the norms, actions and rules to structure, regulate, sustain and hold accountable a sector or set of actors. However, as the humanitarian energy sector is an emerging set of institutions and organisations, governance mechanisms are not as defined as they are in other parts of the humanitarian world. For example, in terms of accountability and rule-making there is no intergovernmental or regulatory body designed to control and monitor humanitarian energy activities. In some circumstances, UNHCR, IOM, and other refugee operators play this role as they usually have responsibility for camps and to some extent regulate which projects and energy activities occur. Responsibilities for firewood and cookstove provision usually lies with UNHCR or the national government in refugee settings, while for internally displaced people IOM takes on this remit. In many contexts the responsibilities for food and fuel are divided – with UNHCR, IOM, and other operators coordinating household fuel needs, and WFP and partners as the main agencies coordinating food-related resources.

In many cases, implementation of firewood distribution is outsourced to a local NGO as an implementing partner and is sometimes overseen by refugee committees or host community agencies. Engagement with host communities and larger-scale energy investments is also overseen to some extent by national governments and governmental ministries. However, this engagement is often largely at the approvals stage, and government officials have limited time to engage. NGOs and private sector actors are also held accountable but mostly to their stakeholders and investing public than to a higher organisational body. The role that different organisations have in structuring and sustaining the humanitarian energy sector is complex and varied. As an increasing number of actors enter the playing field, the nature of governance is likely to change and be challenged. The different types of institutions outlined earlier work together in a range of formal and informal ways, through partnerships, collaborations, cofunding, and sub-contracting.

In general, governance in the humanitarian energy sector can be said to be both global, led by international organisations, and local, led by site offices, implementing NGOs and in-country actors. Relatively few decisions seem to be led by national states or formalised accountabilities within the UN system. Governance so far is informal, multi-institutional, and largely led by certain institutions. For example, while there is no formal cluster for energy within the humanitarian system, the fact that several global policy organisations work together through international forums provides a form of light touch global governance. The Global Plan of Action on Energy for Displaced People, launched in July 2018, is likely to cement these relationships even further.

One of the most common ways of working in the sector is through partnerships and consortiums. Both the Moving Energy Initiative and the SAFE initiative are global partnerships which rely on the members delivering a commitment to work together. This creates an interesting set of relationships and dynamics of collaboration, as the parties are able to work on their own agendas and projects without the involvement of the other partners but choose to try to work together. No one organisation has a core remit on energy for displacement, and working together allows institutions to combine their skills and areas of expertise to make a more comprehensive whole than would have otherwise been possible. Together it is possible for organisations to implement and manage energy projects in refugee camps, whereas as individual organisations they would achieve much less.

Such consortiums and partnerships are common, as are more traditional cofunding and sub-contracting relationships. Contracted relationships have a slightly different mode of governance and organisation, as they usually involve one party sub-contracting the other to do a specific piece of research or project. For example, Project Gaia is an official implementing partner of UNHCR and has been contracted to provide ethanol-fuel stoves in refugee camps in Ethiopia (Project Gaia, 2017). Often these types of relationships are determined by the funding body, in this case UNHCR, who have a specific remit and project they want completed. However, the implementing partner usually has some flexibility in the types of solutions and way the project is managed, and of course can choose not to apply to be an implementing partner or for funding at all.

In summary, the institutional progression of humanitarian energy has been led by a range of actors, not least UNHCR, who are increasingly investing in energy and environmental programming in country offices. There is also a recognition at the global level about the changing nature of the debate and shifting to long-term development planning. As the sector evolves, it is expanding into multiple sites of practice: with policies, programmes, and projects being formed within global discussions, as well as practices being developing at the household and community levels, as well as a great many sites in between.

Conclusion: leaving no one behind and governance in humanitarian energy

The chapter has provided an overview of humanitarian energy issues, including the structures of governance, institutions, and practices involved in energy for displacement. Leaving no one behind in the transition to energy access (SEforALL, 2018) requires displaced people and humanitarian agencies to be supported in their energy needs. Good governance and developing energy skills within institutions is critical to delivering energy access in humanitarian settings. However, as this chapter suggests, there is some way to go before agencies and practitioners can enable access to modern energy products and services across the wide range of displacement settings.

Several things are clear: first, that energy for displacement is a new and emerging sector, one which will require considerable investment to ensure that policies and practices are designed to meet the needs of refugees and displaced people. The importance of inclusive practices is vital in this transition: energy service programmes in displaced contexts are likely to be more successful if organisations implement participatory, bottom-up, and market-based approaches which consider energy technologies for the long-term, displaced people and local communities as agents of change, and use approaches which embed beneficiary agency and market-based solutions (Rosenberg-Jansen et al., 2018).

Second, global progress is being made. Compared to other humanitarian sectors, historical progress has been made quickly. Some of the key lessons which have emerged from the energy access and development sectors, and wider humanitarian aid world have already been embedded at the heart of humanitarian energy action. For example, the sector has quickly embedded market-based approaches and the value of involving the private sector in supplying sustainable solutions (Van Landeghem, 2016).

Finally, the multi-institutional and complex nature of governance and implementation within humanitarian energy suggests that progress at the ground-level may be slow initially, but that intelligent practices can be proposed by non-governmental and research organisations to support real change. The influence of institutions on the humanitarian energy sector is already evident from this short overview. This chapter has highlighted how the scope and definitions with humanitarian energy are continuing to be developed, which institutions are involved in the sector and initial views on how they act, and how projects and activities within refugee camps are arranged.

Bibliography

Bates, E. et al. (2002), 'Participatory Approaches for Alleviating Indoor Air Pollution in Rural Kenyan Kitchens', *Boiling Point*, 48, 12–15.

Bellanca, R. (2014), 'Sustainable Energy Provision Among Displaced Populations Policy and Practice', *Chatham House*, UK.

Birikadde, G., Clancy, J. and Matinga, M. (2009), 'Fuel Security and Supply Dynamics in Internally Displaced Persons' Camps of Northern Uganda', *Journal of Humanitarian Assistance*. http://sites.tufts.edu/jha/archives/462.

Bizzarri, M. and Barbelet, V. (2011), 'Review of Safe Access to Firewood and Alternative Energy (SAFE) Uganda', *World Food Programme*, Rome, Italy.

Bizzarri, M., Bellamy, C., Katajisto, M. and Patrick, P. (2010), 'WRC SAFE Kenya Appraisal Report', *World Food Programme and Women's Refugee Commission.*

Colenbrander, E., Eales, A. and Walley, L. (2018), 'Social Impact of Mini-grids: Monitoring, Evaluation and Learning', *University of Strathclyde*, Scotland, UK.

Coventry University (2017), 'HELP Refugee Project', *University of Coventry*, Coventry, UK.

Egziabher, A. and Murren, J. O'Brien. (2006), 'An Ethanol-fuelled Household Energy Initiative in the Shimelba Refugee Camp, Tigray, Ethiopia: A Joint Study by the UNHCR and the Gaia Association', *Gaia Association*, London, UK.

EUEI PDF (2017), 'About us', *EU Energy Initiative Partnership Dialogue Facility*.

Florini, A. and Sovacool, B. K. (2009), 'Who Governs Energy? The Challenges Facing Global Energy Governance', *Energy Policy*, 37.

Global Alliance for Clean Cookstoves (2013), 'SAFE: Energy Access in Humanitarian Settings', *Global Alliance for Clean Cookstoves*, Washington, DC.

Global Alliance for Clean Cookstoves (2014), 'Global Alliance for Clean Cookstoves: Cooking in Humanitarian Settings', *Global Alliance for Clean Cookstoves*, Washington, DC.

Global Plan of Action for Sustainable Energy Solutions in Situations of Displacement – GPA (2018), *United Nations Institute for Training and Research (UNITAR)*, Geneva, Switzerland.

Grafham, O. and Lahn, G. (2018), 'The Costs of Fuelling Humanitarian Aid', Moving Energy Initiative Report, *Chatham House*, London, UK.

Grafham, O., Lahn, G. and Lehne, J. (2016), 'Energy Solutions with Both Humanitarian and Development Pay-Offs', *Forced Migration Review*, 52, University of Oxford, Oxford, UK.

Grandi, F. (2017) cited in UNHCR (2017), 'Comprehensive Refugee Response Framework', *The Office of the United Nations High Commissioner for Refugees*, Geneva, Switzerland.

Gunning, R. (2014), 'The Current State of Sustainable Energy Provision for Displaced Populations: An Analysis', *Chatham House*, London, UK.

Hilhorst. D., Desportes, I. and de Milliano, C. W. J. (2019), 'Humanitarian Governance and Resilience Building: Ethiopia in Comparative Perspective', *Disasters*, 43 (52).

Humanitarian Response (2018), 'What Is the Cluster Approach?' *United Nations Office for the Coordination of Humanitarian Affairs*, Geneva, Switzerland.

IASC (2009), 'Decision Tree Diagrams on Factors Affecting Choice of Fuel Strategy', *IASC Inter-Agency Standing Committee*, Geneva, Switzerland.

IKEA Foundation (2017), 'IKEA Foundation Brighter Lives for Refugees', *IKEA Foundation*, Amsterdam, Netherlands.

Intermediate Technology Development Group (1996), 'Household Energy in Emergency Situations', *Boiling Point*, 37.

Internal Displacement Monitoring Centre (2015), 'Global Estimates 2015: People Displaced by Disasters', *IDMC*, Geneva, Switzerland.

Internal Displacement Monitoring Centre (2018), 'Global Report on Displacement 2018', *IDMC*, Geneva, Switzerland.

Keating, M. (2015), 'Generating Hope for Refugees', *The World Today*, Chatham House, London, UK.

Lahn, G. and Grafham, O. (2015), 'Heat Light and Power for Refugees: Saving Lives, Reducing Costs', Chatham House report for the Moving Energy Initiative, *Chatham House*, London, UK.

Lehne, J., Blyth, W., Lahn, G., Bazilian, M. and Grafham, O. (2016), 'Energy Services for Refugees and Displaced People', *Energy Strategy Review*, 13–14, 134–146.

Moving Energy Initiative (2017), 'Resources, Moving Energy Initiative', *Chatham House*, London, UK.

Nakicenovic, N., Rogner, M. and Srivastava, L. (2013), 'Towards energy as a sustainable development goal', TERI and IIASA Report.

OCHA (2010), 'Energy Security and Humanitarian Action (No. OCHA Occasional Policy Briefing Series – No. 3)', *United Nations Office for the Coordination of Humanitarian Affairs,* Geneva, Switzerland.

OCHA (2018), 'Cluster Coordination', *United Nations Office for the Coordination of Humanitarian Affairs.*

Practical Action (2018), 'Energy Access: Sustainable Access to Renewable Energy Sources for All', *Practical Action*, Rugby, UK.

Project Gaia (2017), 'Humanitarian work in Ethiopian Refugee Camps', *Project Gaia*, Gettysburg.

Rogers, C., Sovacool, B. K. and Clarke, S. (2013), 'Sweet Nectar of the Gaia: Lessons from Ethiopia's "Project Gaia" ', *Energy for Sustainable Development*, 17.

Rosenberg-Jansen, S. (2018a), 'Energy for Displacement "Understanding the Challenges" ', HEED Briefing Paper No.1, *Scene Connect*, Coventry, UK.

Rosenberg-Jansen, S. (2018b), 'Reflections on the Berlin conference on Energy for Displaced People: A Global Plan of Action', *University of Oxford*, Oxford, UK.

Rosenberg-Jansen, S., Barlow, M., Peisch, S., Ponnan, N. and Rathi, P. (2018), 'Sustainable Humanitarian Energy Services: Inclusive Participation, Lessons Learnt and Paths Forward', Poor Peoples Energy Briefing Series, *Practical Action*, Rugby, UK.

Rowling, M. (2018), 'Global Plan Powers up to Supply Clean Energy for Refugees', *Reuters*.

SAFE (2015) *Safe Access to Fuel and Energy (SAFE)*, Washington, DC.

SAFE Humanitarian Working Group (2017), 'ENERGYCoP: Humanitarian Energy Community of Practice'. http://energycop.safefuelandenergy.org/web/energycop.

SEforALL (2017), 'Taking the Pulse: Understanding Energy Access Market Needs in Five High-Impact Countries', *Sustainable Energy for All*, Vienna, Austria.

SEforALL (2018), 'Sustainable Energy for All (SEforALL)', *Sustainable Energy for All*, Vienna, Austria.

UNDESA (2018), 'SDG 7 – Sustainable Development Goal 7 on Universal Energy Access', *United Nations Department of Economic and Social Affairs (UNDESA)*, New York.

UNHCR (1996), 'UNHCR Environmental Guidelines', *The Office of the United Nations High Commissioner for Refugees*, Geneva, Switzerland.

UNHCR (2014), 'Global Strategy for Safe Access to Fuel and Energy (SAFE). Division of Programme Support and Management', *The Office of the United Nations High Commissioner for Refugees*, Geneva, Switzerland.

UNHCR (2017), 'Comprehensive Refugee Response Framework', *The Office of the United Nations High Commissioner for Refugees*, Geneva, Switzerland.

UNHCR (2019), 'Global Strategy for Sustainable Energy: 2019 to 2024', Geneva, Switzerland.

UNITAR (2018), 'Global Plan of Action on Sustainable Energy Solutions for Situations of Displacement', *United Nations Institute for Training and Research*, Geneva, Switzerland.

Van Landeghem, L. (2016), 'Private-Sector Engagement: The Key to Efficient, Effective Energy Access for Refugees', Toolkit for the Moving Energy Initiative, *Chatham House*, London.

World Bank (2017), 'The Global Tracking Framework Report 2017', *World Bank*, Washington, DC.

2 The migration-energy nexus in international policy

Eva Mach[1]

Global migration figures at a glance

Current estimates suggest that almost 15% of the world's population are on the move. While there were 258 million people living outside of their country of origin in 2017 (IOM, 2017a), a conservative estimate from 2007 suggested that another 740 million moved within their own countries in the early 2000s (UNDP, 2009). If current patterns of migration[2] continue, the number of international migrants could surpass 400 million by 2050 (IOM, 2017a). Internal migration is also expected to continue to grow as part of the global trend towards urbanization. According to the latest United Nations estimates, 68% of the world's population is projected to live in urban areas by 2050, up from 55% in 2018, an increase due to internal migration from rural to urban areas and overall population growth (UNDESA, 2018).

Although most people migrate voluntarily, the current high levels of forced migration[3] remind us that many people might have no choice but to move. According to the United Nations High Commissioner for Refugees (UNHCR), forced migration as a result of persecution, armed conflict or generalized violence was at a record high in 2018 with 70.8 million individuals living in displacement. Out of these people, 25.9 million were refugees[4] residing outside of their countries of origin, 3.5 million seeking asylum, and another 41.3 million were internally displaced people[5] (UNHCR, 2019). Whereas conflicts and violence have led to the internal displacement of on average 7.6 million people per year between 2008 and 2017, disasters have been displacing three times more people, an average of 24.6 million people per year (IDMC, 2018). This trend is likely to continue as the effects of climate change are predicted to intensify and increase the frequency of both extreme weather and climate events (IPCC, 2014). In addition to conflicts and disasters, millions of people have also been displaced due to infrastructural projects. The exact magnitude of such development-induced displacement[6] is currently unknown but it was estimated, for example, in 2006, that around 15 million people per year were forced to migrate due to widespread dam construction taking place across the globe (Cernea, 2006; see for example also: Eguavoen and Tesfai, 2011; Tay, 2017).

Increasingly it is also recognized that the traditional dichotomic labels of forced or voluntary migration are not representative of migration realities on the ground. The concept of mixed migration refers to migration flows where both voluntary and forced migrants travel on the same routes and means of transportation (IOM, 2008). Thus, refugees, asylum seekers, and other migrants might arrive in transit and destination countries together. Mixed migration also acknowledges that when people migrate it might be for a combination of reasons – economic, political, social and environmental (OM, 2008). For example, political drivers, such as protracted conflict, in combination with environmental factors, such as disasters resulting from natural hazards, can together lead to the final decision to migrate. The concept of mixed migration reflects the diverse spectrum of migration itself and the intrinsic challenges faced in migration management and migration data collection.

Understanding the migration-energy nexus in the international policy context

Ever since the Industrial Revolution, energy use and access to energy have been factors shaping human migration patterns. While hydro and steam power increasingly replaced human and animal power, the Industrial Revolution also led to deep structural transformation of societies, contributing to the migration of people from rural areas towards industrial centers and to cities.

Although there is currently a lack of rigorous research and empirical evidence, the nexus between migration and energy represents an emerging field of study at research and policy levels.

The purpose of this chapter is to provide an overview of the research policy context in which energy and migration issues can be analyzed. It does this by examining two conceptual frameworks related to migration governance and management. These are the migration, environment and climate change (MECC) and the migration and development frameworks. Ultimately, with high levels of forced migration and climate change impacts already manifesting worldwide, both approaches represent guiding principles for public policy domains. This paper argues that the migration-energy nexus should be mainstreamed in policies with three goals in mind: helping people to stay by minimizing forced and unmanaged migration as much as possible; helping people on the move by ensuring assistance and protection for those affected while seeking durable solutions; and helping people to move in a safe and orderly manner.

The migration, environment and climate change (MECC) framework

The migration, environment and climate change (MECC) policy discussions have looked at the impact of environmental change on migration patterns, and vice versa, the impact of migration on the environment. Under the MECC nexus, energy security has been considered as a possible environmental factor influencing the decision to migrate. Acknowledging the multi-causality of environmental migration (Figure 2.1), the *Foresight: Migration and Global Environmental*

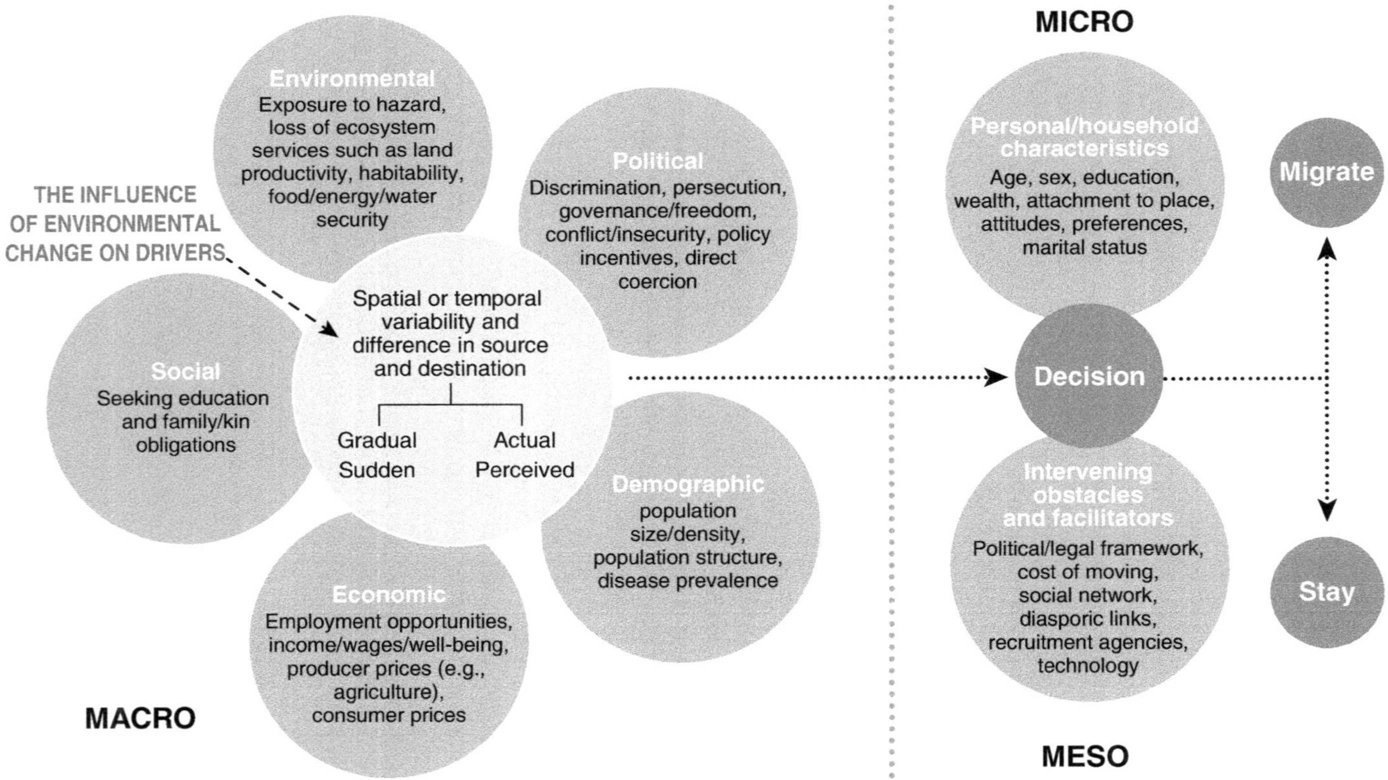

Figure 2.1 Foresight conceptual framework showing the 'drivers' of migration and the influence of environmental change

Source: Foresight, 2011.

Change report noted that energy security played a role in migration (2011). More specifically, the report stated that energy production is one of the "important provisioning services that affect people's decisions to stay or move from their settlements". The report also highlighted the potential inadvertent impacts of climate change mitigation policy on migration, such as the development of large-scale energy projects which might lead to displacement.

The MECC nexus also analyses the impact of migration on the environment, looking at how migration – especially large-scale movement – reshapes the environment in places of destination. In such context, it is crucial to consider energy through the lens of natural resource management, especially when local energy sources and options for harnessing alternative energy are limited.

The migration and development framework

Secondly, the link between migration and energy can be analyzed through a development lens, as both migration and energy are key contributing factors to sustainable development. On the one hand, energy is a basic requirement for economic development and a key contributor to human development. Modern energy services enable people to meet basic human needs and contribute to social development by improving education and public health (Gaye, 2007). Universal access to modern energy services is an indicator of higher economic development with heat, light and mobility fueling economic growth. Sustainable energy in particular has been seen as a key element to support the implementation of the post-2015 agenda for sustainable development as it responds to two development challenges at the same time: driving economic and human development through increasing access to energy, while keeping the commitments related to decreasing greenhouse gas emissions. Related to the nexus, for instance, a case study from Nepal argued that renewable energy technologies enable holistic community development and their mindful implementation could lower the rate of out-migration from rural areas (Zahnd, 2013). The findings of the study suggest that alleviating poverty and bringing development to rural areas through access to affordable, reliable and clean energy services can address some of the drivers of migration.

On the other hand, migration has been a megatrend of the 21st century with around 1 billion people currently on the move. Consequently, human mobility has become a significant national and international concern as well as a powerful engine of development. The importance of migration was acknowledged when, for the first time, the global development framework recognized the integral role of well-managed migration in achieving sustainable development (SDG 10.7). Migration is seen, inter alia, as a means to reduce inequalities and to cope with poverty, local unemployment, loss of livelihoods, environmental degradation and disasters.

One way to show the link between migration and development for policy making purposes is through the *transitional model* (de Haas, 2007), which describes the general effect of development on migration patterns. The model posits that while international migration rises with increased economic development, levels of migration start to decrease after a certain level of development. In short, high overall economic development – for which access to energy is a prerequisite – is associated with reduced emigration.

Some empirical studies support the validity of this model and the assumption that development and poverty reduction through access to energy impact migration patterns. One of these examples can be found in a follow-up study conducted by the Government of Brazil, evaluating the impact of electrification in rural settings. The impact study related to the government program called Programa Luz para Todos found that rural out-migration was reversed after access to modern energy services was improved, with 5% of migrants returning. It also found that 1.7% of the households who had indicated their plans to migrate decided to stay after the arrival of energy services (2013).

Although empirical evidence is in short supply, the development-migration-energy nexus also offers a framework for analysis to better understand how migration might influence access to energy and the energy mix used. For example, a study from Vietnam estimated the potential effects of migration on residential energy use in the capital, where urbanization and migration are occurring at the same time (Komatsu et al., 2013). Komatsu et al found that estimated energy consumption is lower when population increase in urban areas is generated by rural to urban migration, than when urban to urban migration and natural population growth are the main drivers of population increase. Another study focused on the direct energy consumption of rural-to-urban migrants in Beijing (Ru et al., 2015). Ru et al proposed that changes in lifestyles and living conditions led to substantial changes in household energy use as incoming migrants replaced biomass fuel options with coal, electricity and liquefied petroleum gas. By providing empirical evidence on the nexus, such studies have important implications for migration (including forced migration), energy and urbanization policies.

Understanding through these frameworks the migration-energy nexus in the international policy context can help advance the topic on a global scale through prioritizing areas for further research to validate assumptions and identifying evidence-based best practices to mainstream in policies.

Evolution of the migration-energy topic in the global agenda

Even before the Sustainable Development Goal (SDG) 7 was dedicated to universal energy access, development policy recognized the key enabling role of energy services in achieving poverty reduction and sustainable development. The pivotal role of energy – in particular, energy efficiency, energy production and consumption, development of renewable energy sources – was enhanced when environmental concerns were integrated in international development policies, moving beyond a purely economic approach. In that respect, the critical role of energy was recognized in the Brundtland Report (1987) as well as the Rio Declaration (1992) under various sections dedicated to changing consumption patterns, protecting the atmosphere, combating desertification and drought, managing solid waste and sewage and strengthening the role of farmers (UNGA, 1992).

The World Summit on Sustainable Development (WSSD), which took place in Johannesburg in 2002, and the related outcome document, the Johannesburg Plan of Implementation (JPOI), recognized the direct link between energy services and poverty reduction and development. Even though no Millennium Development Goal was at the time dedicated to energy, the JPOI called for improving

> efforts to work together at all levels to improve access to reliable and affordable energy services for sustainable development sufficient to facilitate the achievement of the Millennium development goals, including the goal of halving the proportion of people in poverty by 2015, and as a means to generate other important services that mitigate poverty, bearing in mind that access to energy facilitates the eradication of poverty.
>
> (WSSD, 2002)

On the other hand, the role of migration in sustainable development was not articulated in these frameworks.

Before the adoption of the SDGs, academia, NGOs and international organizations were primarily looking at the migration-energy nexus from the perspective of access to sustainable energy in situations of displacement (see for example Benoit-Xavier, 2011; Franceschi et al., 2014). The topic was increasingly analyzed through quantitative and qualitative interdisciplinary research conducted by academics and think tanks on the field (see for example Lahn and Grafham, 2015; Ossenbrink et al., 2017), while humanitarian organizations started making more and more use of renewable energy solutions to address challenging field conditions, such as the lack of (access to) an electric grid in places of operations. Where implemented for household use, sustainable energy technologies allowed displaced persons living in camps to have better access to lighting, electrification, cooking, heating and cooling and water and sanitation. It can be argued that these technologies – when managed well with a community-based approach in its center – improve the well-being of both displaced people and host communities, contributing to sustainable development beyond the confines of the displaced persons' camps. Introducing such technologies and know-how in displacement settings can help bridge the humanitarian-development divide whereby self-reliance and livelihood diversification are enabled via access to renewable energy.

Interest in understanding the impacts of access to energy on migration patterns rose around the beginning of the so-called European migration crisis in 2015, when growing numbers of migrants entered the European Union by crossing the Mediterranean Sea or overland via Southeast Europe. That growing interest is reflected in the increasing number of events and publications dedicated to the nexus. One of the first political acknowledgement of the importance of the topic came from the Council of the European Union, who recognized "that lack of or uneven access to energy is part of the root causes of irregular migration" (Council of the European Union, 2016, p. 5).

The migration-energy nexus was discussed on several occasions in 2017 and 2018, including at the European Union Energy Initiative Partnership Dialogue

Facility discussion forum: *Enlightening the Migration Debate: the Importance of Sustainable Energy Access* (2017), at the Vienna Energy Forum side event: *Clean Energy for Migrants and Vulnerable Groups* (2017), the Future Energy Forum: *Renewable Energy and Quality of Life* conference (2017), the *Energy for Displaced People: A Global Plan of Action for Sustainable Energy Solutions in Situations of Displacement* conference, the Sustainable Energy for All Forum deep dive session *Energy in Humanitarian Settings* (2018) and *A Global Plan of Action for Sustainable Energy Solutions in Situations of Displacement: Framework for Action* launch conference (2018). Policy makers and practitioners argued in these fora that energy, and renewable energy especially, can act as a stabilizing force that contributes to address some of the drivers of migration (Morales, 2017). In these discussions, sustainable energy also emerged as a stand alone concern in humanitarian work supported by a group of energy and humanitarian practitioners working to accelerate progress towards achieving SDG 7 in forced migration settings through the Global Plan of Action for Sustainable Energy in Displacement Settings.

Alongside these advocacy efforts, a growing number of programs and projects used renewable energy technologies in migration contexts, proving the economic and technical feasibility to use renewable energy in migration management, including in forced migration settings. This portfolio of projects continues to expand, as several development and humanitarian agencies and NGOs include renewable energy components in migration management and humanitarian activities.

The migration-energy nexus in international policy processes

Despite the limited empirical evidence available on the migration-energy nexus, international policy frameworks provide several entry points to consider the topic. Accordingly, this section analyzes how migration and energy have been included in relevant international processes and what this means for policymaking.

Migration and energy in the post-2015 development agenda

The year 2015 represented a watershed year for global policy, with governments committing to bold initiatives aiming to drive lasting change in the development and humanitarian sectors. In March, UN Member States gathered at the Third UN World Conference on Disaster Risk Reduction (WCDRR) in Sendai City, Japan to agree on a new disaster risk reduction agenda, the successor of the Hyogo Framework for Action (HFA). The Sendai Framework for Disaster Risk Reduction 2015–2030 (Sendai Framework) was endorsed subsequently at the General Assembly in June 2015 and established the first major policy framework of the post-2015 development agenda, with seven targets and four priorities for action. While the HFA did not make any reference to energy nor to migration, the Sendai Framework calls for a "people-centered preventive approach to disaster risk", making a direct reference to migrants as actors and contributors to resilience (UNGA, 2015a). Energy is not explicitly mentioned in the Sendai Framework, although, climate-adaptive and disaster risk reduction measures focused on sustainable development are identified as pivotal to implement the Framework. In that respect, considering energy – especially renewables – is required to ensure the implementation of such measures.

In July 2015, UN Member States came together for the United Nations Third International Conference on Financing for Development to showcase their commitments to financially enable the implementation of the post-2015 development agenda. The Addis Ababa Action Agenda (AAAA) paved the way for the adoption of the most inclusive development framework in human history, the 2030 Agenda for Sustainable Development. UN Member States affirmed their "strong political commitment to address the challenge of financing and creating an enabling environment at all levels for sustainable development". Both migration and energy (including energy efficiency) were recognized in the AAAA as essential areas contributing to sustainable development.

In September 2015, the landmark 2030 Agenda for Sustainable Development was adopted at the United Nations Sustainable Development Summit, with the goal to "transform our world" and with the motto to "leave no one behind". The Agenda was called "a plan of action for people, planet and prosperity" (UNGA, 2015b). It is the first time that a global framework for development explicitly recognized the integral role of migration on the one hand, and energy on the other, to contribute to sustainable development. This is in contrast to the Millennium Development Goals, the predecessor to the 2030 Agenda, which contained no direct references to neither migration nor energy.

Under the 2030 Agenda, migration is seen as a means to "reduce inequality within and among countries" and was rooted in SDG 10 under target 10.7, calling to "facilitate orderly, safe, and responsible migration and mobility of people, including through implementation of planned and well-managed migration policies". Beyond target 10.7, the SDGs include numerous references to migration, inter alia, including addressing forced migration like environmental migration (for example, via target 1.5, 2.4, 6.4, 11.5, 11.b, 13.1, 13.2, 13.3), trafficking (via target 16.2) and harnessing "positive contributions of migrants for inclusive growth" via labor mobility and remittances (IOM, 2017a).

Recognizing the importance of energy to power sustainable development by enabling social and economic development, the Agenda dedicated a stand alone goal to energy, as well as making connections to the other 16 goals. SDG 7: *Ensure access to affordable, reliable, sustainable and modern energy for all* is an illustration of the inclusive approach "to leave no one behind", and this broad inclusive understanding provides one of the entry points to consider the migration-energy nexus in the policy context. In line with Goal 10 and target 10.2: *by 2030 empower and promote the social, economic and political inclusion of all irrespective of age, sex, disability, race, ethnicity, origin, religion or economic or other status*, the successful implementation of SDG 7 would ensure that all migrants, including those forced to migrate and/or live in displacement, regardless of their status, enjoy access to modern energy services. It is also possible that this very access to better energy services might lower the risks of distress migration, offering people incentive to remain in their places of origin. For these reasons, the SDG framework, along with the human rights-based approach, provides the best policy framework to develop policy and operational solutions seeking to ensure access to energy to all migrants including displaced persons.

The year of global policy frameworks culminated in Paris in December 2015. The Paris Agreement under the United Nations Framework Convention on Climate Change (UNFCCC) was signed by 197 Parties to the Convention and later ratified by 181 Parties in order to combat climate change both through mitigation and adaptation. The Paris Agreement made specific references to migration and human mobility considering Parties' "respective obligations to . . . migrants" under the Preamble and establishing the Task Force on Displacement "to develop recommendations for integrated approaches to avert, minimize and address displacement related to the adverse impacts of climate change" under COP Decision on Loss and Damage. The Paris Agreement also acknowledged "the need to promote universal access to sustainable energy in developing countries, in particular in Africa, through the enhanced deployment of renewable energy". While no direct reference is made to the migration-energy nexus, the implementation of the Paris Agreement and climate action goals are not possible without shifting the energy mix to renewables in all areas of life – including those in migration settings – and considering the human mobility implications of the low-carbon transformation and climate change impacts.

In October 2016, the New Urban Agenda was adopted at the United Nations Conference on Housing and Sustainable Urban Development (Habitat III) to establish global standards for sustainable urban development (UNGA, 2016c). Both energy and migration occupy preeminent roles in the new global urban policy framework. Migration and migration management are recognized as key issues to consider in building sustainable cities. The Agenda calls for ensuring access to equal opportunities without discrimination and envisions "cities for all, referring to the equal use and enjoyment of cities and human settlements". In line with this call for inclusivity, Member States committed to promote measures to support "equitable and affordable access to sustainable basic physical and social infrastructure for all . . . , including affordable . . . modern and renewable energy". Since cities consume over 60% of the global energy produced and are responsible for 70% of the globe's greenhouse gas emissions (Habitat III, 2018), the New Urban Agenda placed a key emphasis on renewable energy as a measure to ensure environmental sustainability and support cleaner cities. While there is no direct reference in the text to the migration-energy nexus, the Agenda provides a basis for policy development on this topic in urban settings.

Migration and energy in migration and humanitarian policy

While 2015 represented a watershed year for global development policy, 2016 was a catalytic year for humanitarian policy, while setting in motion the development of the first international migration policy framework, the Global Compact for Safe, Orderly and Regular Migration (GCM), as well as a new refugee response framework, the Global Compact for Refugees (GCR).

The first ever World Humanitarian Summit took place in Istanbul in May 2016. The purpose of the Summit was to catalyze global action in the humanitarian sector to deliver better for people suffering in humanitarian crises. The Summit provided a platform to discuss new approaches to manage displacement while ensuring dignity for displaced people and supporting host communities. Forced migration as a way to respond to conflict and disasters constituted a central theme of the Summit with the Agenda for Humanity explicitly calling for reducing displacement and addressing migration under Core Responsibility Three – to leave no one behind (UNGA, 2016a). While energy is not explicitly mentioned in the Agenda for Humanity, the commitments to leave no one behind in situations of conflict and disaster, cannot be sustainably fulfilled without incorporating energy in programming. Relevant anchorage to energy can be found under the displacement and migration calls: to ensure economic sustainability and livelihood generation for displaced people; to establish multi-year frameworks in order "to end aid dependency and promote the self-reliance"; and to ensure long-term political and financial support to host countries and communities, "including for housing, employment, education, health care and other vital public services" (UNGA, 2016a). Beyond Core Responsibility Three, the Agenda and its transformative nature offers a conducive policy environment to advance the migration-energy nexus in the humanitarian sector through the commitments to shift the focus of humanitarian work "from delivering aid to ending need" and to finance the sector differently via "investing in stability and shifting from funding to financing" (UNGA, 2016a). In addition to the Agenda for Humanity, more than 30 of the biggest donors and aid providers committed to changing how the humanitarian sector works through signing up for the Grand Bargain. Signatories pledged inter alia to "increase collaborative humanitarian multi-year planning and funding" and to "enhance engagement between humanitarian and development actors", which are both preconditions for sustainable energy programming in forced migration settings.

Following the Agenda for Humanity, the New York Declaration for Refugees and Migrants was adopted at the UN Summit for Refugees and Migrants in September 2016, illustrating the political will of UN Member States to collectively address migration considerations on a global scale to save lives and protect human rights. Two processes seeking to establish governance frameworks were set in motion by the New York Declaration: the Global Compact for Safe, Orderly and Regular Migration for governing international migration and the Global Compact on Refugees for governing refugee issues (UNGA, 2016b).

Furthermore, the New York Declaration for Refugees and Migrants also gave consideration to the question of the impact of migration on the environment, demonstrating the political will of world leaders to address the environmental impacts and infrastructural challenges linked to large population movements: "85. In addition to meeting direct humanitarian and development needs, we will work to support environmental, social and infrastructural rehabilitation in areas affected by large movements of refugees"; and "6. States, in cooperation with multilateral donors and private-sector partners, as appropriate, would, in coordination with receiving States: . . . e) provide assistance to protect the environment and strengthen infrastructure affected by large movements of refugees in host countries" (UNGA, 2016b).

As a follow up to the New York Declaration and the experiences from the application of the Comprehensive Refugee Response Framework, the development of both the GCM and the GCR offered additional entry points to discuss the migration-energy nexus. In line with the New York Declaration, the Global Compact on Refugees incorporated energy considerations under two elements: financing refugee assistance through maximizing private sector contributions for the "development of innovative technology, including renewable energy, particularly with a view to closing the technology gap and supporting capacity in developing and least developed refugee-hosting countries"; and ensuring efficient natural resource management by addressing "the accommodation and environmental impacts of large numbers of refugees", which would include "contributions . . . to invest in closing the technology gap and scaling-up capacity development for smart, affordable and appropriate technologies and renewable energy in developing and least developed refugee hosting countries". Clean energy is considered one of the ways to address environmental impacts of large population movements, respond to the needs of refugees for reliable energy through "safe access to fuel and energy programming", while providing a hook for better collaboration with the private sector (UNHCR, 2018b).

Although the Global Compact for Safe, Orderly and Regular Migration does not make any direct reference to the migration-energy nexus, there are several ways energy could play a role in supporting the implementation of the GCM. First, the text states clearly in the Preamble that the Compact rests on the purposes and principles of relevant and established policy processes related to environmental protection and sustainable development, including the 2030 Agenda for Sustainable Development, the United Nations Framework Convention on Climate Change, United Nations Convention to Combat Desertification and the Paris Agreement. This also means that the energy, in particular renewable energy, dimensions of these policy processes are de facto of relevance. Second, the GCM also incorporates the set of commitments outlined in the New York Declaration for Refugees and Migrants, including those related to environmental considerations. Furthermore, several of the GCM objectives are of direct relevance to the migration and energy nexus, such as Objective 2 that looks at how to "minimize the adverse drivers and structural factors that compel people to leave their country of origin". Access to sustainable energy and its positive consequences – especially in terms of local development and sustainable livelihood opportunities and the productive use of energy – are critical to address the drivers of migration and contribute to the achievement of the GCM vision of a world where migration is safe, orderly and regular. The text, under Objective 2, also makes clear reference to climate action to accelerate States' fulfilment of the Sustainable Development Goals, especially Sustainable Development Goal 13: *Take urgent action to combat climate change and its impacts*, with the aim of eliminating the adverse drivers and structural factors that force people to leave their countries. While the GCM does not contain any specific reference to access to sustainable energy, it is evident through the indirect references to sustainable development and climate action that clean energy can play a key role in fulfilling the goals of the GCM, in line with the migration and energy commitments under the Agenda 2030.

As outlined previously, direct references to the migration-energy nexus are limited, yet the global policy developments achieved throughout 2015, 2016 and 2018 offer direct and indirect entry points to discuss and advance the topic. Most significantly, the structure of the Agenda 2030, with its stand alone energy goal and its separate migration target, provides the opportunity to consider migration and energy across different development areas. The humanitarian-development nexus approach, which integrates governance structures across sectors, was inherently embedded in these new policy frameworks, and it enables policymakers to address topics even if not explicitly mentioned under SDG goals and targets. Second, the 2030 Agenda focuses on three key enabling factors that allow for consideration of the migration and energy nexus: the importance of cross-sectoral collaborations, the need to strengthen partnerships (SDG 17) and to connect the various policy frameworks together.

Lessons for policy makers

This analysis has shown that research on migration and energy has been scarce, and policy level discussions have tended to be limited to specific migration contexts like forced migration settings and to specific energy solutions such as renewable or clean energy technologies.

Nonetheless, two complementary approaches are gaining prominence at the policy level in addressing the issues related to migration and migration management. The first examines the need to mitigate the impacts of population movements on the environment and to consider the changing landscape of energy patterns. It looks at the "win-win" potential of clean energy in the context of population movements, including in forced migration settings. As such it considers how clean energy solutions can reduce the impact of migration on local natural resources by providing an alternative to biomass and fossil fuel used for energy generating purposes for household consumption as well as for livelihood generation. This approach is also unfolding in the follow-up processes of the Paris Agreement. Discussions on 'just transition' are ongoing under the UNFCCC, which indicates that the topic of migration and energy will remain relevant in global migration and climate change policies. For example, based on previous experiences of development-induced displacement, installing clean energy infrastructure may result in changed migration patterns (see for example Rignall, 2016) or forced migration as people are asked to leave areas slated for large-scale infrastructure development. People might also end up migrating due to structural changes in the society triggered by the transition towards a low-carbon economy; as seen in the past with for instance the closure of coal mines driving migration (see for example Haney and Shkaratan, 2003; Andrews-Speed et al., 2005; Culter, 1999).

Adopting a decarbonized economic model and adapting to the impacts of climate change will have direct impact on people, their livelihoods and their decision to migrate. This is also the reason why the Petersberg Climate Dialogue IX concluded that it is paramount to put "people at the center of climate policy" and to give "due consideration to those whose livelihood and jobs will be affected by structural changes arising from climate policy" (BMU, 2018).

The second approach is rooted in a development perspective and views energy as a stabilizing force that can contribute to address the drivers of migration. For instance, building energy infrastructure – especially renewable energy projects – is one way to create job opportunities that contribute to positive climate action, enhance populations' standards of living, allow to sustain and diversify livelihoods and thus address the drivers of migration. This argument was also put forward by the United Arab Emirates during the third meeting of the preparatory process for the intergovernmental conference to adopt a global compact for safe, orderly and regular migration at the 71st session of the General Assembly (UN Web TV, 2017). It also emerged in the voluntary national reviews (VNR) in 2018, the follow-up and review mechanisms of the Sustainable Development Goals. Out of the 46 VNRs submitted in 2018, 39 countries highlighted migration as a key trend of our time to be managed and a contributor to sustainable development, whereas 43 mentioned energy and renewable energy as a prerequisite for sustainable development. A direct reference to the link of migration and energy was made by the Republic of Benin, considering sustainable energy as a way to address the root cause of rural-urban youth migration (Republic of Benin, 2018).[7]

There is a diverse array of priority actions that policymaking can consider in order to address the migration-energy nexus. As a prerequisite for specific policy interventions, the following takeaway messages are proposed based on the previous analysis:

1 Based on the current state of knowledge, it is not possible to establish direct causal relationships between migration and access to energy. Nor is it possible to establish that access to energy is a sole pull factor of migration or vice versa, that the lack of access to energy acts as a single driver of migration. Empirical research targeting the various aspects of the migration-energy nexus is needed to better understand the links between the two. Enhancing knowledge sharing related to the topic would help to define information gaps, prioritize research needs and identify opportunities for shaping public policies.
2 While it is clear that more empirical research is needed to provide detailed and context-specific evidence, the international community needs to recognize the multiple ways migration and energy interrelate, especially how lack of energy security represents a potential driver of migration; how ensuring energy security is indispensable for upholding human rights throughout the migration cycle including in forced migration settings; and how investing in sustainable energy might help address the drivers of migration in countries of origin.

3 Due to the interdisciplinary nature of the topic, governments need to adopt a whole-of-government approach to address the manifold realities of the migration-energy nexus and consider it across national policies related to development, energy development, migration management, humanitarian preparedness and response, environmental protection and climate action.
4 Policy makers should consider integrating migration in energy policies and vice versa energy in migration governance and management. On the one hand, policy makers working on migration management and humanitarian policy domains should take energy security into account throughout the entire migration cycle: before, during and after migration occurs. On the other hand, policy makers in charge of development, energy, environmental and climate change policies, should consider the challenges and opportunities related to migration at different levels of government.

Reducing the complexity of the migration-energy nexus to certain linkages can dehumanize the discourse and neglect what really happens in people's lives when they are forced to migrate, or the manifold opportunities access to clean energy might bring to enhance the well-being of affected populations. Considering this during the national implementation and the follow-up and review mechanisms of the SDGs, Sendai Framework, New Urban Agenda, Paris Agreement and the Global Compacts is key to address the present and future challenges and opportunities related to the migration-energy nexus – for the benefit of both current and future migrants.

Notes

1 This article represents the view of the author only and not of the International Organization for Migration. The author would like to thank Dina Ionesco (Head of the Migration, Environment and Climate Change Division, IOM) and Mariam Traore Chazalnoel (Migration, Environment and Climate Change Thematic Specialist, IOM) for their contributions.
2 The International Organization for Migration – the UN Migration Agency (IOM) defines a migrant as any person who is moving or has moved across an international border or within a State away from his/her habitual place of residence, regardless of the person's legal status; whether the movement is voluntary or involuntary; what the causes for the movement are; or what the length of the stay is.
3 The International Organization for Migration – the UN Migration Agency (IOM) defines forced migration as a migratory movement in which an element of coercion exists, including threats to life and livelihood, whether arising from natural or man-made causes (e.g., movements of refugees and internally displaced persons as well as people displaced by natural or environmental disasters, chemical or nuclear disasters, famine or development projects).
4 A refugee is a person who, "owing to a well-founded fear of persecution for reasons of race, religion, nationality, membership of a particular social group or political opinions, is outside the country of his nationality and is unable or, owing to such fear, is unwilling to avail himself of the protection of that country." (Art. 1(A)(2), Convention relating to the Status of Refugees, Art. 1A (2), 1951 as modified by the 1967 Protocol).
5 The internationally recognized Guiding Principles on Internal Displacement define IDPs as: persons or groups of persons who have been forced or obliged to fee or to leave their

homes or places of habitual residence, in particular as a result of or in order to avoid the effects of armed conflict, situations of generalized violence, violations of human rights or natural or human-made disasters, and who have not crossed an internationally recognized State border (United Nations, 1998 (E/CN.4/1998/53/Add.2)).

6 Displacement refers to the movement of persons who have been forced or obliged to fee or to leave their homes or places of habitual residence, in particular as a result of or in order to avoid the effects of armed conflict, situations of generalized violence, violations of human rights or natural or human-made disasters. Source: Adapted from Guiding Principles on Internal Displacement, annexed to United Nations Commission on Human Rights, Report of the Representative of the Secretary-General, Mr. Francis M. Deng, Submitted Pursuant to Commission Resolution 1997/39, Addendum (11 February 1998) UN Doc E/CN.4/1998/53/Add.2, 5, para. 2 of the introduction. IOM, Glossary on Migration, IML Series No. 34 (2018), forthcoming.

7 Original text in French: Dans la commune de Bantè, il s'agit de la promotion et l'utilisation des mini-centrales solaires pour la réduction de l'exode rural des jeunes qui, faute d'éclairage, migrent vers les villes environnantes (p. 25).

Bibliography

Akçay, S. and Demirtaş, G. (2015), 'Remittances and Energy Consumption: Evidence from Morocco', *International Migration*, 53 (6), 125–144.

Andrews-Speed, P., Ma, G., Shao, B. and Liao, C. (2005), 'Economic Responses to the Closure of Small-Scale Coal Mines in Chongqing, China', *Resources Policy*, 30 (1), 39–54. www.sciencedirect.com/science/article/abs/pii/S0301420705000036.

Benoit-Xavier, F. (2011), 'Renewable Energy in the Camps of Tamil Nadu', *Forced Migration Review*, 37, 48–49. www.fmreview.org/non-state/Benoit-Xavier.html.

Cernea, M. M. (2006), 'Development-induced and Conflict-Induced IDPs: Bridging the Research Divide', *Forced Migration Review*, Special Issue, 25–27. www.fmreview.org/sites/fmr/files/FMRdownloads/en/FMRpdfs/BrookingsSpecial/15.pdf.

Council of the European Union (2016), 'Outcome of the 3504th Council Meeting, 14969/16, Presse 63, Pr Co 62', *Council of the European Union Press Office*, Brussels. www.consilium.europa.eu/media/22446/st14969en16-v3.pdf.

Culter, S. (1999), 'Managing Decline: Japan's Coal Industry Restructuring and Community Response', *University of Hawaii Press*, Honolulu.

Davies, A. (2012), 'IDPs in Host Families and Host Communities: Assistance for Hosting Arrangements', *UNHCR*, Geneva. www.refworld.org/pdfid/4fe8732c2.pdf.

De Haas, H. (2007), 'Migration and Development: A Theoretical Perspective', *University of Oxford*, Oxford (adapted from Zelinsky (1971), and Martin and Taylor (1996)).

Desai, B. and Anzellini, V. (2018), 'Unsettlement: Urban Displacement in the 21st Century', Internal Displacement Monitoring Centre Thematic Series, *IDMC*, Geneva. www.internal-displacement.org/sites/default/files/inline-files/20180209-idmc-intro-urban-displacement-thematic-series.pdf.

Eguavoen, I. and Tesfai, W. (2011), 'Rebuilding Livelihoods after Dam-Induced Relocation in Koga, Blue Nile Basin, Ethiopia', *University of Bonn*, Bonn. www.econstor.eu/bitstream/10419/88363/1/751992984.pdf.

Fomin and Arc Finance (2014), 'Using Remittances to Finance Clean Energy: Inspiring the Haitian Diaspora to Combat Climate Change', *Fomin*, New York.

Foresight (2011), 'Migration and Global Environmental Change', Final Project Report, *The Government Office for Science*, London. https://assets.publishing.service.gov.uk/government/uploads/system/uploads/attachment_data/file/287717/11-1116-migration-and-global-environmental-change.pdf.

Franceschi, J., Rothkopb, J. and Miller, G. (2014), 'Off-Grid Solar PV Power for Humanitarian Action: From Emergency Communications to Refugee Camp Micro-Grids', *Procedia Engineering*, 78 (2014), 229–235. www.sciencedirect.com/science/article/pii/S1877705814010480.

Gaye, A. (2007), 'Access to Energy and Human Development', Human Development Report 2007/2008, *UNDP*, New York. http://hdr.undp.org/sites/default/files/gaye_amie.pdf.

German Federal Ministry for the Environment, Nature Conservation, and Nuclear Safety (BMU) (2018), 'Co-chairs' conclusions', Petersburg Climate Dialogue IX – Changing Together for a Just Transition, *BMU*, Berlin. www.bmu.de/fileadmin/Daten_BMU/Download_PDF/Klimaschutz/schlussfolgerungen_pcd_ix_en.pdf.

Global Compact for Migration (2018), 'Global Compact for Safe, Orderly and Regular Migration: Intergovernmentally Negotiated and Agreed Outcome', *UN*, New York. www.un.org/pga/72/wp-content/uploads/sites/51/2018/07/180713_Agreed-Outcome_Global-Compact-for-Migration.pdf.

Government of Brazil, Ministério de Minas e Energia (2013), '10 Years, 15 million Beneficiaries: Quantitative Household Evaluation of Satisfaction and Impact Evaluation of the Light for All Program's Main Results', *MME*, Brasilia. www.mme.gov.br/documents/10584/3042878/Pesquisa+2013/c14eec59-80b6-4276-802f56736b3e03df;jsessionid=6C7C37690FDF34D0B537CA9EE0DE52ED.srv155.

Haney, M. and Shkaratan, M. (2003), 'Mine Closure and Its Impact on the Community: Five Years after Mine Closure in Romania, Russia and Ukraine', Policy Research Working Paper 3083. The World Bank, Europe and Central Asia Region, Infrastructure and Energy Services Department, *World Bank*, Washington, DC. http://documents.worldbank.org/curated/en/500791468776712950/Mine-closure-and-its-impact-on-the-community-five-years-after-mine-closure-in-Romania-Russia-and-Ukraine.

Intergovernmental Panel on Climate Change (IPCC) (2014), 'Climate Change 2014 Synthesis Report Summary for Policymakers', Fifth Assessment Report (AR5). *IPCC*, Geneva. www.ipcc.ch/pdf/assessment-report/ar5/syr/AR5_SYR_FINAL_SPM.pdf.

Internal Displacement Monitoring Centre (IDMC) (2018), 'GRID 2018: Global Report on Internal Displacement', *IDMC*, Geneva. www.internal-displacement.org/global-report/grid2018/downloads/2018-GRID.pdf.

International Energy Agency (2016), 'Percentage of Population Relying Upon Traditional Use of Biomass', World Energy Outlook, *IEA*, Paris. Excel Sheet (sent via email and also saved on the shared drive as WEO2016Biomassdatabase-Excel: \\gvafile01\ICP\MPRP_\Environment_Environmental Sustainability\Energy\Research.

International Organization for Migration (IOM) (2011), 'Key Migration Terms', Glossary on Migration Law Series, No. 25, *IOM*, Geneva. www.iom.int/key-migration-terms#Migrant.

International Organization for Migration (IOM) (2017a), 'World Migration Report 2018', *IOM*, Geneva. http://publications.iom.int/system/files/pdf/wmr_2018_en.pdf.

International Organization for Migration (IOM) (2017b), 'Migration in the 2030 Agenda', *IOM*, Geneva. https://publications.iom.int/system/files/pdf/migration_in_the_2030_agenda.pdf.

International Organization for Migration (IOM) (2018a), 'Glossary on Migration', IML Series No. 34, *IOM*, Geneva.

International Organization for Migration (IOM) (2018b), 'Displacement Tracking Matrix', *IOM*, Geneva. https://displacement.iom.int/.

International Organization for Migration (IOM) (2018c), 'Environment and Climate Change in the GCM', Environmental Migration Portal, *IOM*, Geneva. www.environmentalmigration.iom.int/environment-and-climate-change-gcm.

International Organization for Migration (IOM) (2018d), 'UN Summit for Refugees and Migrants', Environmental Migration Portal, *IOM*, Geneva. https://environmentalmigration.iom.int/un-summit-refugees-and-migrants.

Ionesco, D. and Chazalnoel, M. T. (2018), 'Advancing the Global Governance of Climate Migration through the United Nations Framework Convention on Climate Change and the Global Compact on Migration: Perspectives from the International Organization for Migration', Climate Refugees: Beyond the Legal Impasse, *Routledge*, Abingdon, UK.

Kadundu, P., Karhamikire, Le Roux, D., Douzou, S. and Estibals, O. (2016), 'Energy Demand Dynamics beyond Borders: An Exploratory Case Study of Congolese (DRC) Migrants in France', *DEMAND Centre Conference*, Lancaster. www.researchgate.net/publication/320016366_Energy_demand_dynamics_beyond_borders_An_exploratory_case_study_of_Congolese_DRC_migrants_in_France.

Komatsu, C., Dinh Ha, H. and Kaneko, S. (2013), 'The Effects of Internal Migration on Residential Energy Consumption and CO2 Emissions: A Case Study in Hanoi', *Energy for Sustainable Development*, 17 (6), 572–580. www.sciencedirect.com/science/article/pii/S0973082613000860.

Lahn, G. and Grafham, O. (2015), 'Heat, Light and Power for Refugees: Saving Lives, Reducing Costs', Chatham House Report for the Moving Energy Initiative, The Royal Institute of International Affairs, London. www.chathamhouse.org/sites/default/files/publications/research/2015-11-17-heat-light-power-refugees-lahn-grafham-final.pdf.

Mach, E. (2017), 'Implementation of the Migration, Environment and Climate Change-Related Commitments of the 2030 Agenda', Migration in the 2030 Agenda, *IOM*, Geneva. https://environmentalmigration.iom.int/sites/default/files/migration_in_the_2030_agenda.pdf.

Maller, C. (2011), 'Practices Involving Energy and Water Consumption in Migrant Households', Newton, P., *Urban Consumption, CSIRO*, Collingwood.

Manning, D. and Taylor, J. (2014), 'Migration and Fuel use in Rural Mexico', *Ecological Economics*, 102, 126–136.

McCollum, D., Echeverri, L., Busch, S., Pachauri, S., Parkinson, S., Rogelj, J., Krey, V., Minx, J., Nilsson, M., Stevance, A-S. and Riahi, K. (2018), 'Connecting the Sustainable Development Goals by Their Energy Inter-Linkages', *Environmental Research Letters*, 13, 033006.

Morales, H. C. (2017), 'The Role of Sustainable Energy Access in the Migration Debate', *European Union Energy Initiative*, Eschborn. www.euei-pdf.org/sites/default/files/field_publication_file/the_role_of_sustainable_energy_access_in_the_migration_debate_euei_pdf_2017.pdf.

Nordic Development Fund (NDF), Nordic Climate Facility and NEFCO (2015), 'Financing Sustainable Energy Through Remittances Flows to Bolivia', *NDF and NEFCO*, Bolivia.

Ossenbrink, J., Pizzorni, P. and van der Plas, T. (2017), 'Solar PV Systems for Refugee Camps: A Quantitative and Qualitative Assessment of Drivers and Barriers', *ETH Zurich*, Zurich. www.ethz.ch/content/dam/ethz/special-interest/mtec/mtec-department-dam/news/files/solar-pv-in-refugee-camps.

Republic of Benin (2018), 'Voluntary National Contribution to the Implementation of the SDGs at the High Level Political Forum New York, July 2018', Republique du Benin. https://sustainabledevelopment.un.org/content/documents/19366Benin_VNR_2018_BENIN_French.pdf.

Rignall, K. E. (2016), 'Solar Power, State Power, and the Politics of Energy Transition in Pre-Saharan Morocco', *Environment and Planning A: Economy and Space*, 48 (3). http://journals.sagepub.com/doi/abs/10.1177/0308518X15619176?journalCode=epna.

Ru, M., Shu, T., Smith, K., Guofeng, S., Huizhong, S., Ye, H., Han, C., Yilin, C., Xi, C., Junfeng, L., Bengang, L., Xilong, W. and Canfei, H. (2015), 'Direct Energy Consumption Associated Emissions by Rural-to-Urban Migrants in Beijing', *Environmental Science & Technology*, 49 (22), 13708–13715.

Sanchez-Loor, D. A. and Zambrano-Monserrate, M. A. (2015), 'Causality Analysis between Electricity Consumption, Real Gross Domestic Product, Foreign Direct Investment, Human Development and Remittances in Colombia, Ecuador and Mexico', *International Journal of Energy Economics and Policy*, 5 (3), 746–753.

Scott, A. et al. (2018), 'Energy, Migration and the 2030 Agenda for Sustainable Development', *Overseas Development Institute*, London. www.odi.org/sites/odi.org.uk/files/resource-documents/12301.pdf.

Shen, H., Tao, S., Chen, Y., Ciais, P., Güneralp, B., Ru, M., Zhong, Q., Yun, X., Zhu, X, Huang, T., Tao, W., Chen, Y., Li, B., Wang, X., Liu, W., Liu, J. and Zhao, S. (2017), 'Urbanization-Induced Population Migration has Reduced Ambient PM2.5 Concentrations in China', *Science Advances*, 3 (7), e1700300.

Tay, M. (2017), 'Development and Environmental Injustice in Malaysia: A Story of Indigenous Resistance in Sarawak', *EnviroLab Asia*, 1 (1). 1–13. https://scholarship.claremont.edu/envirolabasia/vol1/iss1/7/.

Taylor, M., Moran-Taylor, M., Castellanos, E. and Elias, S. (2011), 'Burning for Sustainability: Biomass Energy, International Migration and the Move to Cleaner Fuels and Cook Stoves in Guatemala', *Annals of the Association of American Geographers*, 101 (4), 918–928.

United Nations Conference on Housing and Sustainable Urban Development (Habitat III) (2018), 'The New Urban Agenda [website]', *UN Habitat*, Nairobi. http://habitat3.org/the-new-urban-agenda/.

United Nations Department of Economic and Social Affairs (UNDESA) (2002), 'Plan of Implementation of the World Summit on Sustainable Development', *UN*.

United Nations Department of Economic and Social Affairs (UNDESA) (2018), 'World Urbanization Prospects: The 2018 Revision, Key facts', *UNDESA*, New York. https://esa.un.org/unpd/wup/Publications/Files/WUP2018-KeyFacts.pdf.

United Nations Department of Economic and Social Affairs (UNDESA), Population Division (2016), 'The World's Cities in 2016: Data Booklet', *UNDESA*, New York. www.un.org/en/development/desa/population/publications/pdf/urbanization/the_worlds_cities_in_2016_data_booklet.pdf.

United Nations Department of Public Information (2018), 'Sustainable Development Goals', Sustainable Development Knowledge Platform, *UNDPI*, New York. https://sustainabledevelopment.un.org/sdgs.

United Nations Development Programme (UNDP) (2009), 'Overcoming Barriers: Human Mobility and Development', Human Development Report, *UNDP*, New York. www.un.org/en/development/desa/population/migration/publications/migrationreport/docs/MigrationReport2017_Highlights.pdf.

United Nations Development Programme (UNDP) (2015), 'Table 1: Human Development Index and its components', Human Development Report, *UNDP*, New York. http://hdr.undp.org/en/composite/HDI.

United Nations General Assembly (UNGA) (1992), 'Rio Declaration on Environment and Development'. www.un.org/documents/ga/conf151/aconf15126-1annex1.htm.

United Nations General Assembly (UNGA) (2015a), 'Sendai Framework for Disaster Risk Reduction 2015–2030', GA Res 69/283, 69th Sess., 92nd Plen. Mtg., Agenda Item 19(c),

UN Doc A/RES/69/283 (23 June 2015). www.un.org/en/development/desa/population/migration/generalassembly/docs/globalcompact/A_RES_69_283.pdf.
United Nations General Assembly (UNGA) (2015b). 'Transforming our World: The 2030 Agenda for Sustainable Development', GA Res 70/1, 70th Sess, 4th Plen. Mtg., Agenda Items 15 and 116, UN Doc A/RES/70/1 (21 October 2015). www.un.org/ga/search/view_doc.asp?symbol=A/RES/70/1&Lang=E.
United Nations General Assembly (UNGA) (2016a), 'Agenda for Humanity', Annex to the Report of the Secretary-General for the World Humanitarian Summit, *UNGA*. www.agendaforhumanity.org/sites/default/files/Secretary-General%27s%20Report%20for%20WHS.pdf.
United Nations General Assembly (UNGA) (2016b), 'New York Declaration for Refugees and Migrants'. www.un.org/en/ga/search/view_doc.asp?symbol=A/RES/71/1.
United Nations General Assembly (UNGA) (2016c) 'The New Urban Agenda'. https://unhabitat.org/wp-content/uploads/2017/05/GA-Resolution-A-RES-71-256-New-Urban-Agenda.pdf.
United Nations Framework Convention on Climate Change (UNFCCC) (2015), 'Paris Agreement', *UNFCCC*, Paris. https://unfccc.int/sites/default/files/english_paris_agreement.pdf.
United Nations High Commissioner for Refugees (UNHCR) (2018a), 'Global Trends: Forced Displacement in 2017', *UNHCR*, Geneva. www.unhcr.org/statistics/unhcrstats/5b27be547/unhcr-global-trends-2017.html.
United Nations High Commissioner for Refugees (UNHCR) (2018b), 'The Global Compact on Refugees: Final Draft', *UNHCR*, Geneva. www.unhcr.org/5b3295167.pdf.
United Nations High Commissioner for Refugees (UNHCR) (2019), 'Global Trends: Forced Displacement in 2018', *UNHCR*, Geneva. www.unhcr.org/5d08d7ee7.pdf.
United Nations International Strategy for Disaster Reduction (2015), 'Sendai Framework for Disaster Risk Reduction 2015–2030', Third World UN Conference, *UNISDR*, Sendai. www.unisdr.org/we/coordinate/sendai-framework.
UN Web TV (2017), '(3rd meeting) Preparatory Process for the Intergovernmental Conference to Adopt a Global Compact for Safe, Orderly and Regular Migration', *General Assembly*, 71st Session. http://webtv.un.org/search/3rd-meeting-preparatory-process-for-the-intergovernmental-conference-to-adopt-a-global-compact-for-safe-orderly-and-regular-migration-general-assembly-71st-session/5445498876001?term=global%20compact.
World Bank (2015), 'Adapting to Higher Energy Costs: Findings from Qualitative Studies in Europeand Central Asia', *World Bank Group*, Washington, DC.
World Bank, International Energy Agency, and the Energy Sector Management Assistance Program (2016), 'Access to Electricity (% of population)', SE4ALL Global Tracking Framework, *World Bank Group*, Washington, DC. https://data.worldbank.org/indicator/EG.ELC.ACCS.ZS.
World Commission on Environment and Development (1987), 'Report of the World Commission on Environment and Development: Our Common Future'. www.un-documents.net/our-common-future.pdf.
World Summit on Sustainable Development (WSSD) (2002), 'Johannesburg Declaration on Sustainable Development and Plan of Implementation of the World Summit on Sustainable Development: The Final Text of Agreements Negotiated by Governments at the World Summit on Sustainable Development', *Johannesburg*, South Africa. www.un.org/esa/sustdev/documents/WSSD_POI_PD/English/WSSD_PlanImpl.pdf.
Zahnd, A. (2013), 'The Role of Renewable Energy Technology in Holistic Community Development', PhD thesis, Murdoch University, Perth. www.springer.com/de/book/9783319039886.

3 Global patterns of forced displacement and energy response

Glada Lahn and Owen Grafham

Introduction

Displacement crises differ widely from region to region and through time, and the success of efforts to improve energy access – as with other welfare interventions – will depend on taking into account local conditions and experiences. While it is true that the majority of refugee or displacement camp residents face acute energy poverty and harsh climates, factors such as the geography, climate, the state of accommodation and facilities, security level and governance of a camp, as well as people's ability and willingness to pay for energy and their energy practices before they arrived to the camp differ. Drawing on recent evidence, this chapter discusses the ways in which such differences will influence the appropriateness of technologies and delivery models as well as the ease of their implementation and prospects for sustainability.

A larger and increasing share of displaced people – an estimated 60% of refugees and 80% of IDPs live among local populations, most in urban areas where their options for access to energy will be tightly linked to their accommodation arrangements and the more general state of access for that city or region. In these situations, it may be not only difficult but particularly detrimental to social cohesion to single out refugees or IDPs for energy access improvement schemes. Turning the question around and asking how better infrastructure and energy services for a neighbourhood, municipality or set of facilities can help build community cohesion and/or help secure tenancy, could be more fruitful. This nascent approach, which is enabling new ways of working in Jordan and Lebanon for example, is likely to gain more traction through the Global Compact on Refugees. It. also brings into the picture those organisations, local and international, who may have been working on sustainable energy access and other related infrastructure prior to or throughout displacement crises, who are not necessarily refugee-focused.

A broad picture of energy deprivation

Camps

Lehne et al. (2016) estimate that the vast majority of displaced people do not have access to enough energy to cover their basic needs (Lehne et al., 2016). This picture is particularly stark in refugee camps where the authors estimate

Table 3.1 Energy deprivation in selected refugee camps

Refugee Camp	*Proportion of sampled population with tier 0 or tier 1 cooking solutions*	*Proportion of sampled population with tier 0 or tier 1 lighting solutions*
Kakuma camp, Kenya	86% tier 0 or tier 1	86% tier 0 or tier 1
Goudoubo camp, Burkina Faso	95% tier 0 or tier 1	67% tier 0 or tier 1
Gihembe camp, Rwanda	85% tier 0 or tier 1	71% tier 0 or tier 1
Kigeme camp, Rwanda	63% tier 0 or tier 1	93% tier 0 or tier 1
Nyabiheke camp, Rwanda	77% tier 0 or tier 1	79% tier 0 or tier 1

Source: Adapted from Corbyn and Vianello (2018) and Sandwell et al. (2019 forthcoming).

that 89% only have tier 0 lighting and an estimated 77% only have tier 0 cooking facilities. Their analysis suggests that, globally, over 7 million displaced people in camps have access to electricity for less than four hours and inadequate access to non-electric energy every day (Lehne et al., 2016). Table 3.1 draws on some more recent granular research which backs up this assessment of widespread deprivation in refugee camps:

Where such levels of deprivation exist, so it appears is the corresponding desire for energy products and services. For example, a recent survey of forcibly displaced Rohingya in refugee camps in Bangladesh found that those who felt aid does not cover their needs (473 people) considered energy to be their most important unmet need – placing energy as the service most people considered unmet (55%) ahead of cash (52%), food assistance (33%) and water, sanitation and hygiene (29%) (Hetzer and Hopkins, 2018).

But these broad trends mask a range of differences between the way that energy is used and provided. For example, although most survey work that has been undertaken in refugee camps in Sub-Saharan Africa highlights the fact that people do place a high priority on cooking and lighting interventions (Okello, 2015; Vianello, 2015; Corbyn and Vianello, 2018) new findings continue to point towards significant contextual differences between specific settings. In Burkina Faso, the large-scale free distribution of solar lanterns has contributed to the higher proportion of camp residents with access to decent lighting solutions. This would deter companies from including the camp in a business plan for further scale up of basic lighting solutions (Whitehouse, 2019), although not necessarily for the sale of solar home systems which could power other devices.

In Tanzania, Rivoal and Haselip found that refugees in Nyarugusu camp were spending as much as 50% of their income on fuel with which to cook whereas surveys of other camps have suggested a lower percentage (10–15% of income – See Okello, 2015; Vianello, 2015) – a variation which would have a significant impact on the type of cooking interventions that would be possible. The cost of wood and charcoal – as with other fuels and services – is a key consideration across the board when considering the economics of substitution, as well as the fact that much is not paid for in cash but in food supplies or in the time (and risk) that women take in collecting it. Such contextual differences should have a significant impact on humanitarian agencies since they show that there is no 'one-size-fits-all' approach to humanitarian energy programming even in settings that one might think of as relatively similar.

Moreover, there is a broad spectrum of what counts as a displacement or refugee 'camp'. Some of the most vulnerable people do not live in officially administered refugee camps but in informal settlements (for example, Syrian refugees in Lebanon) or are caught in transit (Libya, Greece, Sudan, etc.). This further affects what is legally or physically possible in terms of welfare interventions in general. There are also large displaced communities living in camps that have turned into small cities or slums of concrete and wires over the course of several generations (for example, the Palestinian camps in Gaza, Lebanon and Jordan and the Bihari camps in Bangladesh). These, like most well-established camps, will have developed and entrenched coping mechanisms and economies for the buying and selling of power and fuel, however unequal and unreliable supplies may be. As with the urban situations discussed in the following section, improving energy conditions must take into account the need for urban upgrading and network resilience.

Urban displacement

Outside of the camp environment, the differences in energy needs between displaced groups become more pronounced. Around 60% of refugees and 80% of IDPs find refuge among the host population, largely in urban areas (Muggah and Abdenur, 2018) with a wide variety of local energy conditions. Lehne et al. estimate that in urban, slum and rural settings, the distribution of those with tier 0 access to energy is not quite as extreme as it is for those in camps (Lehne et al., 2016). They suggest that around 14% of urban forcibly displaced people have tier 0 lighting and an estimated 9% have tier 0 cooking facilities. These figures sound relatively low, but these lower proportions still represent around 4 million and 2.6 million displaced people for lighting and cooking respectively.

Even when people have greater access to energy services and products, this does not mean that they do not face significant pressures and problems. Where IDPs and refugees are living in slums or inadequate housing, maintaining decent sanitation and healthy temperature and humidity levels may be a constant challenge and stressor on family life. As with life in camps, finding enough money to pay for fuel, hot water and power will often mean a reduction in ability to spend on other essentials. These needs are interlinked and bound up with housing and local and municipal infrastructure. Where IDPs and refugees are living in rented accommodation, shared energy and water services may mean that bill payments for utilities become a point of tension between tenants and landlords (Lahn et al., 2016).

Living outside of camps may make displaced groups harder to reach – it may be more difficult to identify who is most in need and less clear how to improve their energy access. Interventions that single out displaced groups for assistance may reinforce negative perceptions of unfairness among the host community.

In a pioneering approach, the Norwegian Refugee Council's Shelter Programme in Jordan has tried to avoid this by conducting building upgrades and energy improvements to secure tenancy for refugees by reducing bills and improving the building's value for the landlord. This required building the relationships and the legal contracts to ensure that the arrangements work for both parties and are legally sound after the sentence: (See chapter 4).

A few aid programmes have begun to address energy as part of decent housing projects in countries suffering from large-scale displacement and energy access issues. There is a strong case for this, even while it may not target refugees or IDPs specifically, as housing issues are often an acute point of tension following a displacement crisis. For example, funded by French development agency (AFD), the NGO GERES has worked in the context of rapid and unplanned urbanization in Kabul to trial 'bioclimatic solutions for urban housing' using insulation and solar energy.[1] The UK Department for International Development-funded Moving Energy Initiative supported partners Jordan Green Building Council and Habitat for Humanity Jordan to conduct energy and water efficiency retrofits for 48 low-income homes and help three families to build new green affordable homes. Such projects also aim at growing local skills, including those of refugees, and developing local markets for sustainable energy and building materials.

Similarities, differences and interactions with host populations

Aid to refugee crises and the structure of the humanitarian system has tended to encourage the treatment of refugees as distinct group of 'beneficiaries', to be addressed separately or in isolation from hosting communities. This view carried into energy projects may mean local or national governments prevent projects from being implemented altogether (if they do not benefit the host community) or miss opportunities for scale – for example by widening the market to the local area.

A shared lack of energy access

In some cases, when it comes to energy, the actual differences in energy access between refugees and the surrounding host communities can be relatively small. For example, the limited energy access that refugees have in Kakuma refugee camp in Kenya should not be surprising in the context of the wider energy access issues in the area. Kenya has a rapidly increasing overall electricity access rate (grid and off-grid) of 75% (2018) according to the World Bank's Multi-Tier Framework Energy Access Survey Report. However, this figure falls to 60% in rural areas, and just 5% in Turkana West, which hosts Kakuma refugee camp and the Kalobeyei integrated settlement (UNHCR, 2018).

For both the population of Kakuma camp, and the surrounding host community, most cooking is done with biomass (primarily wood) on basic cookstoves or three-stone fires. Almost none have access to electricity (despite the fact that Kakuma town is connected to the diesel mini-grid (Patel, 2019). In many countries around the world, host communities are almost as marginalised and impoverished as the refugees or displaced people around them. In many cases, refugees come from situations where the energy access level was similar or even lower than the one that they find themselves in. In a few cases, refugee camp dwellers have greater access to energy products than those in the surrounding community (Corbyn and Vianello, 2018). Where the overall energy access problems of forcibly displaced people resembles the energy access problems of the host community, there is a case for addressing both at the same time (Grafham and Vianello, 2018).

Starker contrasts between displaced people and host communities have been experienced in Europe, where hundreds of thousands of people fleeing conflict began to arrive by sea, mainly to Greece and Italy from 2015. In the Aegean islands of Greece, reception centres turned camps have seen over 1.13 million arrivals between the beginning of 2015 and mid 2019, most of whom are trying to reach other countries (IOM, 2019). The energy conditions (including lack of heating and hot water in freezing conditions) in camps including Moria on Lesvos and Vathi on Samos and Nea Kavala refugee camp have remained inhumane (Lahn et al., 2016). Tourism and services in nearby areas have been badly hit by the crisis and while many locals have and continue to try to provide relief where they can, relations with host communities have been tense, with some instances of violence as well as peaceful protests over poor conditions. Although most locals have far higher levels of energy and water access, some towns like Mytiline on Lesvos suffer frequent electricity outages and have been paying higher than necessary costs for additional power from oil-fuelled ships (Lahn et al., 2016).

Proximity to urban areas

In Kenya, only 13% of refugees reside in urban areas, with the majority of the remainder split between Dadaab and Kakuma camps. This means that engagement with the host community most often takes the form of interacting with those local populations in the vicinity of the camps in question, through local markets and public facilities (many facilities such as hospitals in the camp serve both the refugee and host community), as well as through specific enterprises, businesses and traders. The proximity of particular camps to urban centres, villages and the grid will shape the services that camp inhabitants can access, and their relative cost and affordability, thereby presenting different contexts for energy access interventions. As the BBOXX study in Chapter 11 notes, the cost of logistics for bringing the company's solar home systems to sell to inhabitants of Kakuma town and refugee camp have been "50% higher than in the case of other areas where BBOXX have shops, largely due to the lack of existing traditional infrastructure and the remoteness of the location".

Reliance on municipal services

With conditions in camps so dire, an increasing number of people try to move to the cities or to informal settlements close to urban areas. In contrast to the situation in many sub-Saharan African camps, refugees from the Syrian crisis beginning in 2012, are not predominantly housed in camps. In Jordan, under 20% of the Syrian refugee population live in the official camps. The remainder are resident, mainly in urban settings, among the host community. As a result, direct engagement between host and refugee populations occurs on a daily basis as they access the same public facilities and markets and often live next door to one another. In such situations, the interactions between the forcibly displaced and the humanitarian community is more limited. In these circumstances, cities and host governments become the primary service provider, and energy services are likely to be under pressure alongside other municipal services (water, waste, health, education, policing, etc.). Here, it makes more sense for approaches to energy to be focused on strengthening the resilience of local services, infrastructures and supply chains.

For humanitarians and development actors, this could mean examining the ways in which energy and infrastructure improvements might support the areas in which they are already working in non-camp areas of a country, such as shelter, health, education or WASH. This theme is developed in Chapter 4 by Jourdi, Gharaibeh and Qahoush, who discuss emerging work on addressing energy for public buildings.

Governance issues

The level of political direction at the national (and even regional) level regarding the displacement challenge, the level of decentralisation and the level and effectiveness of coordination at the host country level and between agencies, NGOs and other actors will all influence the feasibility of energy access projects. Governance issues will be the overarching consideration for all interventions and deserve in-depth attention that we do not have space for here. However, a few cases give an idea of the range of governance contexts and issues that would affect planning.

In Greece, for example, the lack of wider European coordination on resettlement of people seeking refuge, the lack of vision for Greece's long-term resilience in view of its status as a transit country and the lack of government coordination severely hampers installation of improved basic facilities for camps (Lahn et al., 2016). Confusion about who needs to give authority for interventions, and contention over responsibility for camp management have been two of the many issues that have dogged attempts to improve conditions, including for winter heating. One site that has deployed solar energy is Kara Tepe in Lesvos, which has been overseen not by a mix of national authorities and international agencies but the island's mayor (Lahn and Rantsiou, 2017).

By contrast, in Uganda, the government has allowed large (predominantly South Sudanese) refugee settlements on its territory to expand and assume permanent characteristics (Haselip, 2018). The allocation of land for cultivation by refugees enables livelihood projects linked to food, water and energy such as small-scale agroforestry. One of the largest refugee camp complexes in the world, Bidi Bidi, is scheduled to become a city, which aims to integrate both Ugandan and refugee populations (Strochlic, 2019). The Ugandan Rural Electrification Agency plans to extend the grid to reach the northern part of the camp, while other development focused agencies are planning mini-grid options for other parts.[2]

The status of local or national government policy and plans relating to energy are also important considerations. For example, does it make sense to establish a solar-home systems business or wait for the government to extend the grid as part of an existing national energy access plan? And could potential bans on woodfuel use affect cooking projects (as has been proposed in Tanzania and Rwanda)?

Short-term and long-term outlooks for energy provision

Two reasons are often given for why the refugee- and displacement-focused humanitarian agencies cannot or should not consider energy more systematically. The first is that humanitarian response is a short-term mode of delivery, and doing energy properly is a long-term project. The second – related – argument is that in these short-term settings, energy cannot be considered as important as other needs such as shelter, food, water and sanitation and so forth.

Timeframes are a key consideration since displacement crises will require different approaches at different levels, depending on their nature and age. Any given situation may also be subject to change at short notice, for example through the sudden closure of a camp due to security concerns. This uncertainty presents many challenges for anyone seeking to undertake energy interventions.

In short-term emergency response, the consideration that humanitarian agencies give to energy products as essential equipment for distribution is growing. Stoves to cook food have been part of non-food-item (NFI) kits for a long time, but it has been rare for agencies to consider the fuel with which displaced people cook beyond rations for wood and/or charcoal to the most vulnerable. This is changing, and agencies also now consider other energy products such as solar lanterns and/or phone charging equipment. The 2018 edition of the Sphere handbook, the fourth edition of the guide which sets out the minimum standards for humanitarian practitioners, goes significantly beyond previous editions in laying out standards for food *and fuel* and also highlights critical considerations and decision-points regarding other energy products, for example solar lamps (The Sphere Project, 2018).

Longer term, to the authors' knowledge, there is at the time of writing no guidelines or strategy in any humanitarian agency for how they deal with energy in protracted situations. Most refugee situations continue well beyond the emergency phase. The average age of a refugee camp, for example, is eighteen years (Grafham and Lahn, 2018). Whilst the humanitarian sector is based on short-term relief of six months to two years, the reality of most people's lived experience of conflict-induced displacement is significantly lengthier.[3]

The problem is that emergency forms of relief, which may include shelters and other camp structures, energy products, diesel generators and the means of water delivery often influence and lock-in consumption over many years and require regular maintenance (Lahn and Grafham, 2015). In Chapter 6 of this collection, Bellanca argues that the humanitarian sector's reliance on the distribution of energy products such as lanterns and cookstoves, has stunted the development of local energy markets and hindered greater uptake of more effective solutions. By measuring success in terms of 'the number of products distributed' rather than the ultimate impact of that solution, Bellanca argues that the humanitarian sector has missed opportunities to scale up demand. Cross et al. (2019) also show that the preconception of displaced people as having a reliance on handouts, obscures the many ways in which refugees are constantly reusing, trading and bartering for energy services.

It would be logical for humanitarian actors to anticipate protraction and at a certain point introduce or hand over to local authorities or development agencies to install or oversee more durable, sustainable solutions. Yet there are two critical barriers: 1) the budgets of humanitarian agencies are normally short-term (most often one year) and are often unable to adapt to new solutions which might require greater capital investments; and 2) governments may not be in favour of infrastructure and economic solutions which suggest or encourage permanence (Lahn and Grafham, 2015).

With regard to overcoming these barriers, some ways through the impasse are emerging. First, greater attention to the problem is generating data that shows that the costs of reliance on emergency measures such as firewood and diesel are damaging other development priorities. These are costs both for the humanitarian agencies and for the hosting country. The example of Box 3.1 shows one way in which short-term thinking around provision of energy supplies can generate negative long-term outcomes, both for displaced people and the environment. Grafham and Lahn (2018) show the huge opportunity cost for humanitarian agencies of reliance on diesel where there are opportunities for cheaper efficiency plus solar solutions. The studies underpinning the Jordan Response Plan calculate the cost of the Syrian refugee population to government fuel, electricity and water subsidy bill (see Box 2 in Chapter 8). Increasing studies are drawing attention to the shared interests between humanitarian and government partners in using longer-term infrastructure approaches to reduce such bills and environmental stress.

Box 3.1 The 'true costs' of relying on the short-term distribution of firewood in Nyarugusu refugee camp, Tanzania.

The real costs of reliance on short-term emergency energy solutions in humanitarian settings are becoming increasingly clear. For instance, in most Sub-Saharan refugee camps, a humanitarian agency will distribute a certain amount of firewood to families in the camp. Sometimes this firewood is directed to all households, and sometimes this is directed to those identified as most vulnerable. In almost all cases, the supply provided by the agency is not sufficient to cover the entire needs of the refugee household but is instead justified on the basis of what can be afforded in the given financial year. In Nyarugusu refugee camp in Tanzania, Rivoal and Haselip highlighted a range of perverse outcomes derived from the overwhelming reliance of the refugee population on firewood. These outcomes included the negative effects on human health as a result of air pollution; the additional time used cooking and time used collecting wood; the amount of money spent on purchasing additional wood supplies; and perhaps most pressingly of all – the local (and sometimes regional) deforestation caused through such widespread and unsustainably managed practices of firewood collection (Rivoal and Haselip, 2017) – which often generated conflict with local communities also reliant on the same resources. Rivoal and Haselip (2017) estimate the economic benefits of avoided forest degradation amount to $371,178 per year, for Nyarugusu, but other benefits can also be derived in terms of human health, time saved for productive activities and in terms of emissions savings. Many other countries also face similar issues, and Lahn and Grafham have previously estimated that an estimated 64,700 acres of forest are burned each year by forcibly displaced families living in camps who have no alternative fuel source (Lahn and Grafham, 2015).

Second, in several countries (e.g., Jordan, Uganda, Kenya, Bangladesh), response and resilience plans agreed between governments and aid agencies from 2015 onwards are showing that energy (and energy-related issues such as forests, transport and decent housing) are national development priorities. These plans enable the improvement in welfare and outlook for refugees and displaced people by making sure that aid also delivers for nationals. As Mach points out in Chapter 2, the Global Compact's wording on for example, renewable energy as an opportunity to close the technology gap and provide local jobs, could pave the way for a more coordinated long-term approach to energy in some hosting countries.

The Compact also paves the way for more creative partnerships both across the sector and beyond it. For example, while the ICRC like others remains highly dependent on diesel for power generation, its longer-term infrastructure focused approach and in-country technical knowledge offers some possibilities for energy- and water-resilience partnerships which join up the facility-community-national level thinking. Interest in supporting sustainable energy in displacement crises from multilateral development banks such as the European Bank of Reconstruction and Development, the World Bank and the Islamic

Development Bank may offer potential to introduce some of the larger-scale infrastructure projects and technical assistance that could link up support for humanitarian projects.

To achieve this vision requires a more coordinated approach between stakeholders than is the case in 2019, and one which learns from and builds on the experiences of response and resilience over the last five years.

Integrating energy through humanitarian practice

As well as introducing energy priorities into agendas and practice in situations of human displacement, our research suggests that humanitarian agencies should also redress their understanding of energy, thinking beyond the simplistic view of 'cooking and lighting' or 'fuel and electricity supplies'. Energy should not be viewed as a one-dimensional, supply-led, addition to basic services (i.e., through a cookstove or a solar lantern), but it should also be understood from an infrastructure and services perspective, as part of an interconnected web of vital services which can better serve humanitarian objectives. The example of water-pumping solutions in Nyarugusu refugee camp is highlighted in Box 3.2 as an example of this.

Box 3.2 Providing water-pumping solutions in Nyaragusu Refugee camp, Tanzania.

Nyarugusu refugee camp is the largest refugee camp in Tanzania. It hosts around 150,000 refugees, primarily from Burundi. The camp has been in operation since 1996, but, like many camps in Sub-Saharan Africa, refugees there have very limited access to energy products or services, and recent fieldwork in the camp suggests that many people there are keen for greater diversity of energy products and services (Rivoal and Haselip, 2017). However, Fohgrub (2018) has shown that the camp is overwhelmingly dependent on diesel generators in order to power the essential services of the camp. At the time of the sample in 2017, 33 diesel generators used 460,000 litres of diesel per year for electricity at a cost of $345,000, emitting 1,260 tonnes of Co2 per year (Fohgrub, 2018). Four water pumping stations in the camp were estimated to be responsible for roughly 60% of the diesel budget (with the hospital, schools, health posts, admin buildings and living compounds using considerably less electricity). The assessment carried out by Fohgrub concluded that introducing a turnkey PV-diesel hybrid solution for the water pumping infrastructure would pay back in less than four years (Fohgrub, 2018).

In the example of Nyarugusu, introducing a hybrid solar solution could save humanitarian agencies money, reduce the negative environmental footprint of existing delivery models and improve overall service delivery.

Beyond the humanitarian sector, there are a range of actors – development agencies and NGOs, local NGOs, companies, multilateral development banks, foundations and impact investors that take an interest in or are actively engaged in long-term sustainable energy work in hosting countries. Response and resilience plans and the Global Compact provide some basis for cooperation to plan evolution from emergency relief to longer-term sustainability in the areas of energy, water and associated infrastructure. Who owns infrastructure, who is responsible for its safety, maintenance and end-of-life management are key questions for projects involving solar facilities, mini-grids, grid infrastructure and metering.

Conclusion

As energy in displacement settings attracts growing attention and becomes the focus of strategies and targets in the humanitarian sector, we must be careful that our responses are not confined to the camp setting of flimsy shelters and 'beneficiaries'. The range of situations and complexity involved across the world demands a contextual approach to energy, whilst there will be successful models that can and should be adapted. The growing body of studies on energy practices, spending and perceptions and case studies on experience should be pooled to help to guide and improve future planning.

Humanitarian agencies who want to move towards more sustainable energy practices have felt hamstrung by the dual constraints of budget and host country political acceptability. The new emphasis on 'energy and infrastructure' in the UN's Global Compact on Refugees as a means of fairer burden sharing and ways in which international assistance can help hosting countries to cope with new populations, offers a way forward through the impasse.

In protracted situations, energy access and services – like water – become more a question of infrastructure and markets than products. The lines between the needs of host and displaced communities blur. How can living quarters be upgraded to provide more liveable temperatures? How can wiring be made safer while instituting a more reliable service? How might a locality provide the economic case for the development of an energy services market? Can new solar installations strengthen the resilience of the national grid? How might an energy transformation of a community building or hospital improve health outcomes? How could solar and remote sensing improve the quality of water supply and enable bill payment across an underserved area? How might a private-public partnership take over the responsibility for service delivery from a humanitarian agency? How could an employment-generating agroforestry project be integrated with the scale up of safer, cleaner cooking? These are sample questions that could engender more locally and politically acceptable – and therefore more durable – solutions.

Notes

1 GERES (Group for the Environment, Renewable Energy and Solidarity) "Dissemination of bioclimatic solutions for urban housing in Afghanistan".
2 See for example, The Smart Communities Coalition (2018) 'Market Profile: Bidi Settlement', 21 December 2018.
3 An important exception in the humanitarian field is the International Committee of the Red Cross (ICRC), which has tended to develop long-running relationships with national utilities (water and electricity) providers in conflict-prone countries as well as having more autonomy over investment of budgets.

Bibliography

Baczynska, G. (2016), 'European Rights Watchdog Complains about Greek Migrant Camps Conditions', *Reuters*.

Corbyn, D. and Vianello, M. (2018), 'Prices, Products and Priorities: Meeting Refugees' Energy Needs in Burkina Faso and Kenya', *Chatham House*, UK.

Cross, J. et al. (2019), 'Stories of Displacement in Eight Objects', *Chatham House*, London, UK.

Fohgrub, T. (2018), 'Energiewende in Refugee Camps: Electricity Needs of Refugee Camps – Case Study Nyarugusu, Tanzania', Presentation given to the Berlin conference of the Global Plan of Action for Sustainable Energy in Situations of Displacement, *United Nations Institute for Training and Research (UNITAR)*.

Grafham, O. and Lahn, G. (2018), 'The Costs of Fuelling Humanitarian Aid', *Chatham House*, UK.

Grafham, O. and Vianello, M. (2018), 'Energy for Displacement: Key Issues', Cited in 'Energy for Displacement 'Understanding the Challenges'' HEED Briefing Paper No.1, Humanitarian Engineering and Energy for Displacement Project, *Scene Connect.*

Haselip, J. (2018), 'Does Uganda's Refugee Policy Enable Sustainable Energy Production and Consumption?', *Moving Energy Initiative*, London, UK.

Hetzer, R. and Hopkins, K. (2018), 'Bulletin – Rohingya: Needs and Services', *Ground Truth Solutions*, Vienna, Austria.

IOM (2019), 'Flow Monitoring Europe', *International Organization for Migration*, Switzerland.

Lahn, G. and Grafham, O. (2015), Heat, Light and Power for Refugees, *Chatham House*, UK.

Lahn, G., Grafham, O. and Sparr, A. (2016), 'Refugees and Energy Resilience in Jordan', *Chatham House*, London, UK.

Lahn, G., Lampert, S. and Grafham, O. (2016), 'Service Delivery in the Refugee Camps of Greece: Improving Coordination and Strengthening Resilience', *Chatham House*, London, UK.

Lahn, G. and Rantsiou, F. (2017), 'Making Refugee Aid Add Up in Greece', Expert Comment, *Chatham House*, London, UK.

Lehne, J., Blyth, W., Lahn, G., Bazilian, M. and Grafham, O. (2016), 'Energy Services for Refugees and Displaced People', *Energy Strategy Review*, 13–14, 134–146.

Muggah, R. and Abdenur, A. E. (2018), 'Refugees and the City the Twenty-First-Century Front Line', World Refugee Council Research Paper No. 2 – July 2018, *Centre for International Governance Innovation.*

Okello, S. (ed.) (2015), 'The Energy Situation in the Dadaab and Goudoubo Refugee Camps', *Chatham House*, London, UK.

Patel, L. (2019), 'Infrastructure Management Contracts: Improving Energy Asset Management in Displacement Settings', *Chatham House*, London, UK.
Rivoal, M. and Haselip, J. (2017), 'The True Cost of Using Traditional Fuels in a Humanitarian Setting: Case Study of the Nyarugusu Refugee Camp, Kigoma Region, Tanzania', UNEP DTU Working Paper Series 2017: 3, *UNEP DTU Partnership,* Technical University of Denmark.
Sandwell et al. (2019), 'A Quantitative Assessment of Refugee Energy Needs and Priorities in Rwandan Refugee Camps', *Practical Action*, Rugby, UK.
Save the Children (2015), 'Multi-Sector Needs Assessment of Refugees and Migrants in Greece', *Save the Children*, London, UK.
Sphere Project (2018), 'Sphere Handbook (2018)', *Humanitarian Charter and Minimum Standards in Disaster Response, 2018.*
Strochlic, N. (2019), 'In Uganda, a Unique Urban Experiment is Under Way', *National Geographic.*
Vianello, M. (ed.) (2015), 'The Energy Situation in Goudoubo Refugee Camp, Burkina Faso', *Chatham House*, London, UK.
UNHCR (2018),'Kalobeyei Integrated Socioeconomic Development Plan in Turkana West', *The Office of the High Commissioner for Refugees*, Geneva, Switzerland.
Whitehouse, K. (2019), 'Adopting a Market-based Approach to Boost Energy Access in Displaced Contexts', *Chatham House*, London, UK.

Part 2

Evolving approaches

4 Energy in the response to an urban refugee crisis

The case of solarizing public buildings in Jordan

Elias Jourdi, Lama Gharaibeh and Shada Qahoush

Introduction

The outpouring of over 5.1 million people to neighbouring countries (mainly Lebanon, Turkey, Jordan, Iraq, and Egypt) since the beginning of the conflict in Syria in 2011, is generating a change in the way that the humanitarian sector, host governments, development actors, and foreign donors respond. The majority of these refugees are concentrated in urban areas where municipal services were under pressure prior to the population increase. Immediate relief efforts have given way to more long-term plans to support security of stay for refugees and sustainable development for the hosting countries. The governments of Jordan and Lebanon have highlighted energy as an area they see as vital to address as part of the refugee response. This agreement has underpinned two of the first ever solar PV plants for refugee camps (in Zaatari and Azraq) but what about urban areas? Can sustainable energy approaches make a difference here too, and what might be the challenges for a humanitarian agency engaging in this nontraditional area?

To answer these questions, we explain Jordan's refugee crisis as an 'urban problem' and consider local and international responses in the area of sustainable energy that the response plan and resilience focus have opened up. We then make the case that better energy services for public buildings can bring mutual benefit for both national and refugee users and discuss in more detail Norwegian Refugee Council's experience with solarizing 23 state schools in the Northern Governorates. This experience highlights some of the challenges that now need to be addressed in the Jordanian context and which may also have relevance for public buildings approaches elsewhere.

Jordan's refugee crisis as an urban problem

Jordan hosts over 662,000 (UNHCR, 2019) registered Syrian refugees – and several other refugee groups including Palestinian, Iraqi, Yemeni, and Sudanese,[1] and has one of the highest refugee-to-host population ratios in the world (see Figure 4.1).

The overall population of Jordan is estimated to have increased by over one-third in a period of just four years (2011–2015) (Government of Jordan Department of Statistics, 2018). And this has been and remains a largely urban phenomenon.

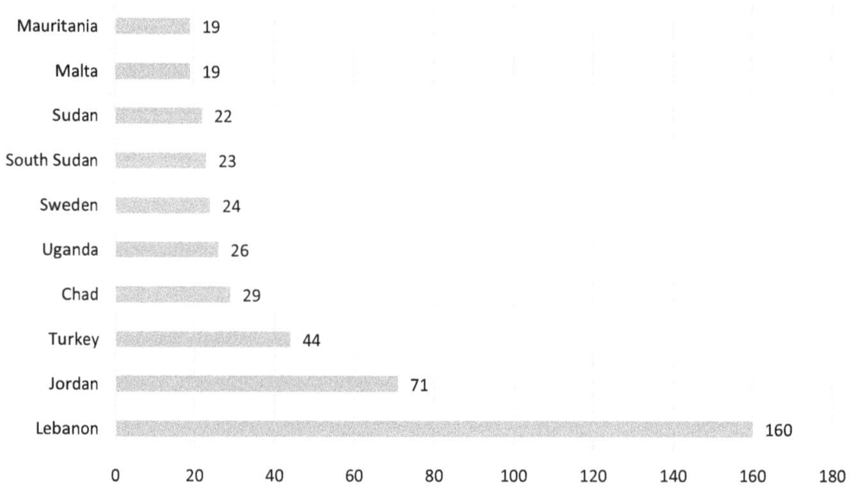

Figure 4.1 Number of refugees per 1,000 inhabitants mid 2018

Source: UNHCR, 2019a.

Over 80 percent (UNHCR, 2019b) of Syrian refugees find accommodation – mostly rented – in cities and towns while the rest live in refugee camps – mainly managed by the UN Refugee Agency (UNHCR).[2] The Northern Governorates and Amman were most clearly affected. According to the 2015 Housing and Population Census (HPC), refugees made up as much as 30 percent of the population of some of the districts in Irbid and Mafraq (Alshoubaki and Harris, 2018).[3]

Cities, in particular, have had to accommodate the rapid population increases. In the cities of Mafraq and Ramtha, the population is said to have more than doubled between 2011 and 2015.[4] In 2019, the majority of Syrian refugees were based in the governorate of Amman, making up almost 30 percent of the population there (UNHCR, 2019b). See Figure 4.2.

This influx of refugees put significant pressure on the country's already struggling economy and, as this chapter will show, caused a rapid increase in demand for municipal services such as health, education, sanitation, and utilities, particularly in the Northern Governorates. Local authorities struggled to cope with the increased demand for solid waste management, water, and electricity, while other sectors including education and housing witnessed increased competition over the already limited resources (MOPIC, 2017). This and the attendant inflation in rental prices have increased social tension (MOPIC, 2017). The use of resources also has implications for refugee tenure security. For example, NRC's shelter and legal aid programmes found that utility bills – chiefly the accumulation of debt relating to rent and utilities bills – were among the most common reasons for conflict between landlords and Syrian tenants (Norwegian Refugee Council, 2016).

The response to the Syria crisis in Jordan has stimulated one of the most carefully examined studies of the economic and social impact of refugees in history,

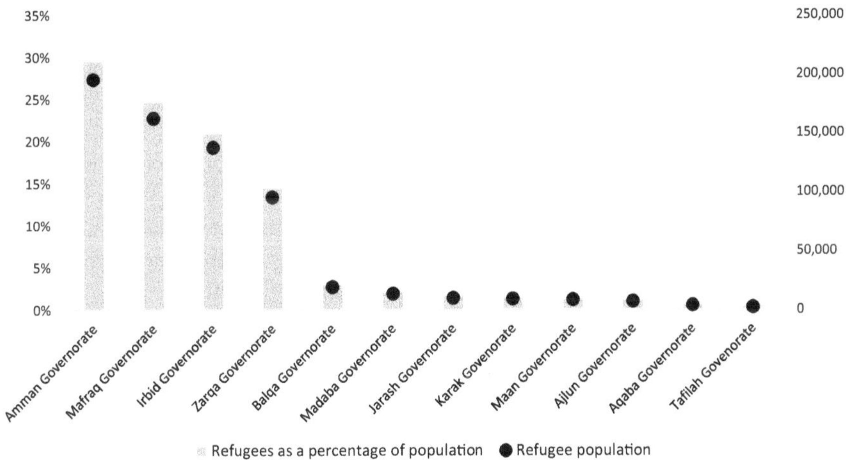

Figure 4.2 Refugee numbers by governorate and as a share of population, mid-year 2019

Source: UNHCR, 2019b.

with data critical to international aid and the 'Jordan Compact', which looks beyond aid. The Jordan Response Plan (see Box 4.1) – a rolling plan that began in 2015 to channel aid to both immediate refugee needs and longer-term country development is becoming a model for how to deal with protracted displacement as a national, and overwhelmingly urban, challenge (Huang and Gough, 2019).

Box 4.1 Energy in the Jordan Response Plan (JRP)

The Jordan Response Plan for the Syria Crisis is a government-led, rolling three-year strategy, with partnership between UN agencies, donors, and local and international organizations for the response to the Syria crisis in Jordan. The plan seeks to address the needs and vulnerabilities of Syrian refugees and Jordanian people, communities, and institutions affected by the crisis. It also focuses on providing a comprehensive vulnerability assessment which combines refugees and national resilience together and sets out one single plan for each of the 7 sectors (these were originally 12): Health, Education, Public services (Local governance and municipal services, transportation, energy and environment), Economic empowerment, WASH, Protection and Justice, and Basic Needs and Shelter (WANA, Chatham House & NRC, 2019). Each of the sectors is managed by a Task Force that combines members from humanitarian agencies local and international and led by a governmental representative from; the JRP helped the government to create partnerships with humanitarian agencies to benefit from the humanitarian experience and engage them in achieving the governmental plans for each sector, since it is the only national document providing for international grants for the Syria crisis.

The costs of energy are unsustainable

Given the pressure on utilities networks and the costs involved (see Box 4.2), it is not surprising that energy – a service that, unlike priorities such as Shelter, Education, and WASH, has no institutional home in the humanitarian sector – has become a priority of the Jordan Response Plan. In its most recent version, 2018–2020, it included an appeal for (172) million USD (MOPIC, 2017) in support of the energy sector over the three years, including a range of projects addressing both refugees and host communities' needs (resilience), indicating a need for financial support of (26) million USD to the energy sector as in the 2019 report [these numbers do not reflect what has been secured, which is far less] (MOPIC, 2019).

The overall objective of the energy chapter of the plan is "to scale up and accelerate responses to Jordan's Energy crisis in a sustainable manner that all alleviate international demand pressures from the Syria crisis" (MOPIC, 2017). Projects were then organized under two specific objectives, the first being to introduce and promote renewable energy and energy efficiency technologies, while the second focused on access to adequate, affordable, and secure energy supply.

Box 4.2 The costs of energy in the residential and the public sectors

Looking closer at the impact of the refugee crisis on the energy sector, in addition to the deterioration of the quality of service and network, the total residential electricity consumption increased by 34 percent between 2011 and 2016 (MOPIC, 2017). Findings from the Department of Statistic's (DOS) 2013 Household Income and Expenditure Survey show that low-income households, which account for a third of the population, are paying a larger share of their total annual expenditures on fuel and lighting compared to any other population group (DOS, 2013). In addition, the government subsidies are the largest for low-income households and can reach up to 81 percent for the lowest consumption brackets. This has increased the government subsidy, as most Syrian households will remain within the subsidized lower electricity consumption blocks.

Fuel for cooking and heating in the form of bottled gas or kerosene is also subsidized. For example, in winter months a family of five would consume between two to five LPG cylinders per month for heating and cooking at an average cost of (9.8) USD per cylinder. The GoJ estimated that subsidies to Syrian households added a further JoD 160 million to its costs in 2017, USD

476 million in water, which includes a subsidy for electricity in the water sector (MOPIC, 2016, p. 171).

According to the General Budget Department, provisions of electricity for public buildings increased by (42%) and fuels increased by (29%) between 2011–2013 (General Budget Department of Jordan, 2014). As an example, the GOJ support for refugees' access to education has led the Ministry of Education to have 98 schools operate double shift classes (Jordanian children in the morning and Syrian children in the evening), to accommodate the influx of (225,000) Syrian children. The ministry increased the number of schools to (208) in 2017 and extended the schools' operation time from five to six days a week (MOPIC, 2017). In addition to the increase in the operational and personnel costs related to the double shift schools' set up, the MOE expenditure on electricity for the primary public schools has as much as tripled between 2013 and 2017, reaching 4.27 million USD (General budget department of Jordan, 2018).

Figure 4.3 demonstrates the high costs of electricity and fuel to the Jordanian civil service. Those ministries administrating public buildings exhibit particularly high bills. For example, in 2017, the total energy bill for the Ministry of Health was 25.6 million JoD (USD 36m) and for the Ministry of Education, 6.8 million JoD (USD 9.6m).

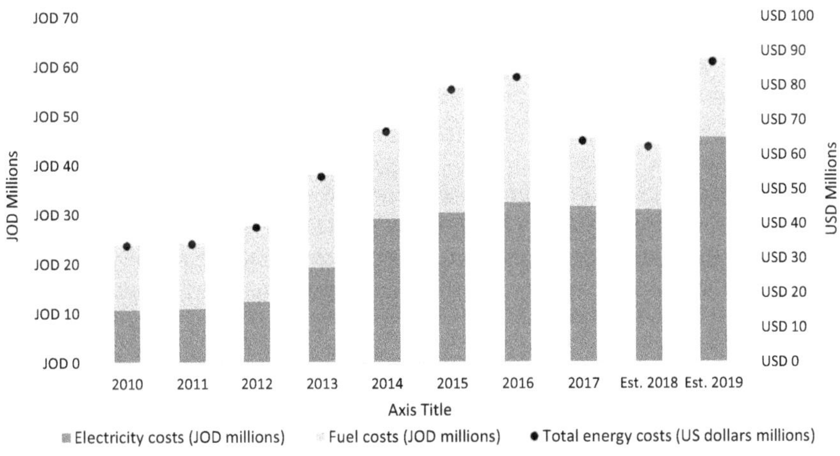

Figure 4.3 Civil service electricity and fuel costs, 2010–2019

Source: Jordan General Budget Law for the Fiscal Year 2019 – Annual reports.

Humanitarian responses in the 'new' area of energy

Jordan provides a relatively positive context for sustainable energy deployment and investments, given its progressive regulation, renewable energy, and energy efficiency targets and developed private sector capacity (see for example Lahn et al., 2016).

Humanitarian projects in the domain of sustainable energy are a relatively new phenomenon, but the energy challenges that organizations have faced – such as high electricity bills for the camps and the demarcation of energy as a priority in the JRP – has encouraged more funding in this area and some creative partnerships.

The focus of projects implemented by humanitarian agencies has to date been mostly on refugee camps, households, and public buildings. These projects included equipping the two largest formal camps of Azraq and Zaatari with solar power plants at each location, partially covering the camp residents' electricity needs. There have been nascent attempts to improve host community-use facilities and households with energy efficient solutions and renewable energy products – something that the Norwegian Refugee Council (NRC) has been pursuing – and these are set to expand. In addition, in 2016, the Jordan Renewable Energy and Energy Efficiency Fund (JREEEF) which sits under the Ministry of Energy, launched an EU backed intervention supporting Jordanian families to cover the cost of installing solar water heaters, directly granting half of the needed amount (the second half is paid by the family). The nationwide program aimed at doubling the number of households using solar water heaters by 2020 around 30 percent as targetted at the NEEAP.

However, the response has not been proportional to the scale of the crisis, as described by the JRP 2018–2020; the needs continue to be huge; there is a need to scale up interventions. This document focuses more on the resilience part where percent of the funds are targeted for supporting public buildings and households and the rest for refugees mainly in the camps to support the grid infrastructure and increase the storage capacities.

This approach to energy in a refugee crisis can be aligned with the country's 'sustainable energy transition', 'resilient cities', and/or 'green growth' agendas, benefiting from the information, studies, experience, and skills already available. This opens up opportunities for humanitarian aid and expertise to contribute to and enhance existing initiatives in a way that seeks to protect the most vulnerable and serve humanitarian principles. A focus on state-owned schools and hospitals, for example, could bring dual benefits for energy sustainability and education and health.

Better energy services for public buildings can improve vital services for all

Public buildings such as government offices and state-owned hospitals, clinics, and schools in Jordan account for 6 percent (NEPCO, 2016) of the total built-up area in the country, with a total number of 107,671 public facilities across the country. The public sector is facing challenges that hinder its abilities to cover the needs and continue providing quality service to users. These challenges impact the day-to-day operation of public facilities. Examples include hospitals limiting their use of water heating for part of the year, which affects the quality of the required hygiene for the hospital and comfort of the patients and state-run schools without heating for classrooms in Jordan's cold winters or proper insulation from humidity or dampness which in turn affects the comfort of the learning environment of the students and teachers and leads to pupil absences according to Ministry of Education (MoE) representatives.

During NRC's technical assessment of schools, we noted that these facilities are meant to be used for generations to come, yet the majority were designed and built without consideration for energy efficiency or a healthy environment for learning. Buildings are generally not insulated, most do not take advantage of natural lighting and ventilation. This results in high operational costs due to the inefficient consumption of fuels for heating, due to energy losses, which increases the yearly maintenance requirement because of the humidity problems. For example, the head of one of the schools in Jerash, which we included in our programme, has to ask for extra support from the MoE each year to address this problem.

Existing public buildings are in urgent need of retrofitting to increase their energy efficiency and reduce their energy bills. The royal initiative for heating public schools, launched in Mafraq and Maan, is an inspiring example of the national movement to install heating using solar systems alongside enhancing buildings efficiency in state-run schools. A pilot by the Ministry of Public Works and Housing in partnership with the German development bank KfW to install solar PV and improve efficiency in five buildings, including two hospitals, a healthcare centre, and a school, was completed in 2019 and is preparing for scale up (WANA Institute, 2019).

However, there are several issues that need to be addressed to get the full potential from this kind of urban response, some of which are explored later through NRC's experience of solarizing schools in Jordan.

Solarizing state schools: NRC's experience

Between 2015–2017,[5] the Norwegian Refugee Council (NRC) supported 23 refugee-hosting public schools with on-grid photovoltaic systems, a project aimed at reducing refugee-hosting public schools' electricity bills with as much as (70–90%). While a scale up of this project is planned,[6] and more projects are planned to target public buildings, there remain many challenges that limit the outcomes of renewable energy projects on public buildings. In our experience, these are:

1 **The sector's technical and operational ability to maintain renewable energy systems following installation and handover**. While the majority of decision makers in the public sector are aware of the importance of maintenance in renewable energy projects in securing their long-term effectiveness, it is not clear where that responsibility sits and if there is the sufficient

investment in the sector's technical capacity. As part of its project, NRC facilitated a training for the engineers of the MOE in Irbid and Jerash where a total capacity of 426 KWP of on-grid PV systems were installed. The training was intended to increase the technical capacity in the MOE. Following the handover of the systems, a few of the trained engineers or the technicians would regularly visit the schools to ensure that the systems were operating efficiently.

With the potential to scale up projects on public schools, hundreds of buildings could benefit from the installation of solar systems and would require a much more significant investment in the ministry's renewable energy technical and operational capacity. The MOE has increasingly faced this challenge and following numerous advocacy efforts, the MOE established the Renewable Energy Unit under the Central Buildings Department in Amman in May 2018. The unit's key role would be to monitor the handed over PV projects on schools across the country. Other alternatives could be through a centralized role for public entities, such as JREEEF and the Samra Electric Power Company (SEPCO), to support public buildings or for municipalities to build expertise in this area or the expertise for contracting out maintenance to the private sector building on the GOJ's decentralisation policy.

2 **The complexity of the approvals processes when implementing renewable energy projects on public buildings**. The approvals process can take several months, and the lack of coordination between the various stakeholders leads in some cases to the duplication of efforts or stalled projects. An example of the lack of coordination is how two governmental initiatives targeted the same schools one with PV roof-top on grid systems and the other with wheeling (whereby the offsite solar generation that is netted off the consumer's bill). Processes for the design and approvals phase are not unified between ministries, nor lessons learned, absorbed, and acted on. This leads to unnecessary delays to the delivery of the project, wasting valuable mobilization resources.

3 **The lack of data**. Data to help design projects is either unavailable or outdated. A significant time and relative resources are therefore initially invested on collecting and analyzing data. Now that a number of entities are working on schools, and national entities such as the National Energy Research Center and JREEEF are accumulating data, there is an opportunity for greater transparency that will improve the quality and value of future projects.

4 **Limited opportunity for scale up or facilities gain through reinvestment of savings**. One of the objectives of the NRC project was to support the creation of a revolving fund managed by the MoE, using the savings from the electricity bills due to the PV systems to support maintenance works for schools. We found that the solar PV systems covered 70 to 90 percent of schools' electricity usage, therefore generating savings for the Ministry, which is responsible for the electricity bills for all state schools. The installed systems are estimated to reduce the bill by around 173,000 USD per year. We estimated that reusing those savings would over a period of seven years be

sufficient to cover all schools in the Irbid district with PV systems. However, the MoE was not able to isolate those savings because of fiscal limitations and the ministry's budget balance. In fact, any savings made by the MOE are not necessarily resources that the ministry is able to reallocate, or even use. The MOE is currently reviewing how to move forward its renewable energy strategy and has identified, following successful advocacy, that the reinvestment of savings from renewable energy projects on public schools has significant potential to achieve better RE outcomes for the ministry. Discussions with MOE representatives highlighted the potential for the establishment of the ministry's RE fund or complementing the JREEEF to overcome the challenge around savings. However, the ministry's budgetary limitations and lengthy governmental processes suggest that such solutions may take years to implement.

The way forward

Finding solutions for the challenges to renewable energy projects in public building will be key to scale up these interventions. The current national momentum on the drive towards renewable energy and energy efficiency practices is an opportunity to create an enabling environment for innovative projects to be implemented on public buildings with a longer-term perspective. Stronger coordination, improved technical capacity, and a data-driven plan, would significantly contribute towards reducing the financial burden of public building energy inefficiencies while improving the quality of the service provided.

The country's National Energy Efficiency Action Plan is a key policy tool, which identifies actions and initiatives to rapidly reduce the energy demand growth. The plan includes promoting and adapting the use of renewable energy and energy efficiency actions to different users. Working in public buildings supports the idea of promoting these kinds of interventions to the local community by raising awareness in adapting these practices.

Finding qualified technical capacities in the governmental facilities for the support of energy projects is hard, and partnership with the private sector can be a solution, since as mentioned earlier the country has the needed expertise that is willing to develop projects for the long term. Such partnernerships can evolve with the right financial models. Private companies could be offered a guarantee for providing EE and RE solutions to public schools and hospitals to cover electricity and fuel costs. Companies would agree to reduce the bill costs by a certain percentage on a monthly or quarterly basis, and the ministries would pay the cost of operation by payments within a certain duration. This service would ensure the O&M for a long period of time from the private sector part.

Furthermore, it is vital to jointly work with the private sector and financial institutions on developing financial models that can enhance the ownership of the public buildings and identifying intervention modalities beyond the transfer of grants that can add financial value to the project. In order to develop

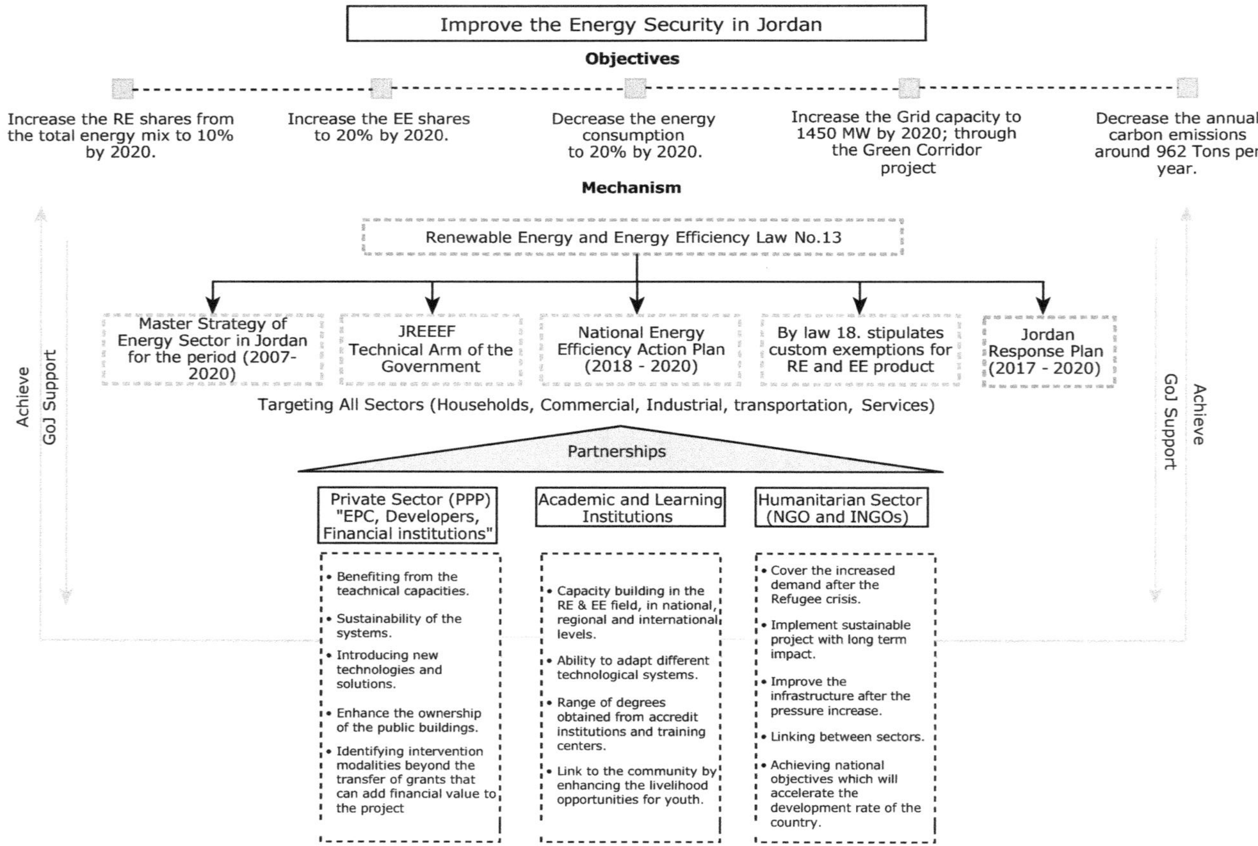

Figure 4.4 Jordan energy security plan

Source: Jordan national energy efficiency strategy, 2018–2020.

these models, it is essential to involve the private sector as it is key to develop long-term, commercially viable solutions. At the same time, projects in this area should help link public entities such as electricity utilities companies and JREEEF and/or work with the municipalities to ensure harmony and long-term sustainability for the systems, especially post-installation, when the guarantee period has expired.

Conclusion

This chapter has considered energy as a way of helping to support humanitarian aims in a protracted, primarily urban displacement crisis. From a humanitarian perspective, energy can be understood as a way to help ease host country pressures and contribute to core humanitarian priorities such as sanitation, health, and education – which are also likely to be local development priorities. In an urban context, public facilities offer an entry point for donors, humanitarians, and local sustainability and development actors to join forces. They will often be under increased pressure to meet the new demand for services, and the burden of additional costs for electricity – as well as fuel and water – may eat into already constrained budgets for health, education, and other social services.

As many humanitarian agencies already collaborate with and assist schools and clinics, for example, clean energy approaches to these buildings and how they can help increase the quality of provision, should be a consideration. These can also align with other initiatives in a given sector to promote scale up. As we have shown, there is huge potential for solarization and efficiency applied at scale in public buildings in Jordan, and this is likely to be the case in many other countries with urban refugee and displacement crises. However, NRC's project to solarize 23 schools in the Northern Governorates encountered several barriers to effectiveness and scale up such as sustainability of maintenance, bureaucratic approvals processes, lack of data, and lack of mechanisms for reinvestment of savings. Our experience demonstrates the importance of working with local and national authorities, the private sector, and other actors in this space to help overcome these barriers.

Notes

1 These include around 370,000 Palestinians living in ten camps, which have become small towns, UNRWA (2019), 'Where We Work'.
2 There is also the Mrajeeb Al Fhood refugee camp of around 4,000 people in Zarqa run by the Emirates Red Crescent and at least one smaller camp under local control.
3 Data adapted from Alshoubaki and Harris, 2018, who estimate "the percentage of Syrians to Jordanians" in 2015 at 37 percent in Mafraq, 19 percent in Irbid, and 11 percent in Amman.
4 Quote from the Mayor of Al-Mafraq city in Malkawi, K. 'Mafraq, Ramtha population doubled since start of Syrian crisis', (*Malkawi, 2015*).
5 Funded by the EU Delegation, as part of REEE1 pilot project in Jordan.
6 Funded by the IKEA Foundation, as part of the Renewable Energy for Refugees (RE4R) project.

Bibliography

Alshoubaki, W. and Harris, M. (2018), 'The Impact of Syrian Refugees on Jordan: A Framework for Analysis', *Journal of International Studies*, 11 (2), 154–179.

DOS (2013), 'Household Expenditures & Income Survey 2013–2014', *Department of Statistics*, Government of Jordan.

DOS (2018), 'Statistical Yearbook of Jordan', *Department of Statistics*, Government of Jordan.

General Budget Department of Jordan (2014), '2011 and 2013 Annual Budget Report', *General Budget Department*, Jordan.

General Budget Department of Jordan (2018), 'Annual Budget Report', *General Budget Department, Ministry of Education*, Jordan.

Huang, C. and Gough, K. (2019), 'The Jordan Compact: Three Years on, Where Do We Stand?' Centre for Global Development Report, 11 March.

Lahn, G., Grafham, O. and El-Sayed Sparr, A. (2016), 'Refugees and Energy Resilience in Jordan, Research Paper for the Moving Energy Initiative', *Royal Institute of International Affairs*, London.

Malkawi, K. (2015), 'Mafraq, Ramtha Population Doubled Since Start of Syrian Crisis', *Jordan Times*, 27 November.

MOPIC (2016), 'Jordan Response Plan for the Syria Crisis 2016–2018', *Ministry of Planning and International Cooperation*, Jordan.

MOPIC (2017), 'JRP 2017–2020 Final Report', *Ministry of Planning and International Cooperation*, Jordan.

MOPIC (2019), 'JRP 2019 Plan Final Report', *Ministry of Planning and International Cooperation*, Jordan.

NEPCO (2016), 'Annual Report 2017', *National Electric Power Company*, Jordan.

NRC (2016), 'SWH Project Field Assessment Data'. *Norwegian Refugee Council*, Internal Resource, Oslo, Norway.

Office of the Prime Minister (2018), 'Jordan's Second National Energy Efficiency Strategy, 2018–2020', *Office of the Prime Minister*, Government of Jordan.

UNHCR (2019a), 'Mid-Year Trends 2018', *Office of the United Nations High Commissioner for Refugees*, January.

UNHCR (2019b), 'Syria Regional Refugee Response-Jordan', *Office of the United Nations High Commissioner for Refugees*, last updated 4 July. https://data2.unhcr.org/en/situations/syria/location/36 (accessed 16 July 2019).

UNRWA (2019), 'Where We Work', *United Nations Relief and Works Agency for Palestine Refugees in the Near East (UNRWA)*, Amman, Jordan.

WANA Institute and Chatham House (2018), 'Expert Roundtable on Scaling up Sustainable Energy for Public Buildings in Jordan', 27 June, *West Asia North Africa Institute*, Amman, Jordan.

WANA Institute, Chatham House and Norwegian Refugee Council (2019), '2nd Meeting of the Inter-Ministry Committee on Sustainable Energy Scale-Up', Renewable Energy for Refugees Project, 2 October, Amman, Jordan.

5 Towards community energy resilience

Long Seng To and Niraj Subedi

Energy resilience and community resilience concepts

The term resilience has been used in many different contexts and fields of study as a way to understand and navigate complexity and uncertainty (Grove, 2018). With its origins in ecology, resilience can be defined as the ability for relationships within the system to persist and absorb changes (Holling, 1973). That is, resilience is not simply the ability to come back to an equilibrium after a disturbance. Social resilience is the "ability of groups or communities to cope with external stresses and disturbances as a result of social, political and environmental change" (Adger, 2000). It is helpful to analyse these aspects of resilience together as socio-ecological systems due to the strong interlinkages between ecological and social resilience, especially where communities are reliant on natural resources for their livelihoods (Berkes et al., 2008).

Research on resilience in terms of energy infrastructure has focused on the engineering perspective (Francis and Bekera, 2014; Panteli and Mancarella, 2015). It has focused on large-scale, critical infrastructure in urban settings operating under high uncertainty (e.g., Roege et al., 2014; Sharifi and Yamagata, 2016). Considerable work has also been undertaken on how critical infrastructures are linked, including modelling of cascading failures (Ouyang, 2014). There is an emerging perspective that infrastructure needs to be flexible, able to be repurposed and redeployed according to changing circumstances, whether the disruptions are sudden or ongoing. There is a need for local innovation to prevent, prepare, respond and recover from disasters. However, traditional models of energy planning provide results that are inappropriate for fragile contexts (Bazilian and Chattopadhyay, 2016).

Energy security is often defined at the national level in terms of the four A's – availability, affordability, accessibility and acceptability (Kruyt et al., 2009). More recently, there has been a shift in emphasis from reducing exposure to risks to building resilience. Cherp and Jewell offer a conceptualisation of energy security as "low vulnerability of vital systems" (Cherp and Jewell, 2014). This approach, focused on vulnerabilities, links vital energy systems and critical social functions, thus linking energy security with human security (Herington and Malakar, 2016). This definition moves the debate beyond security of energy supply to sustained access to energy services where social, economic and technological factors are difficult to predict.

Vulnerability is a combination of exposure to risks and the inability to adapt (Gallopín, 2006). Thus, resilience is closely linked to a community's adaptive capacity. Energy is especially relevant to adaptations to climate change and the management of natural resources. The Intergovernmental Panel on Climate Change defines adaptive capacity as "the ability of systems, institutions, humans and other organisms to adjust to potential damage, to take advantage of opportunities, or to respond to consequences" (IPCC, 2014).

A recent conceptual review of energy resilience and the adaptive capacity of energy systems identified diversity of energy sources, redundancy in the system and organisation structure as important factors (Molyneaux et al., 2016). Renewables have a role to play in increasing energy resilience and reducing vulnerability as they can increase the diversity of energy sources, increase redundancy in the energy system and increase the share of local resources (Spyrou et al., 2019). Localised, distributed approaches to energy can address vulnerability and build resilience (O'Brien and Hope, 2010) but need to be balanced with the greater technical and economic efficiency that more centralised structures can bring (Molyneaux et al., 2016). Distributed energy technologies, such as renewables, have the potential to be more resilient but more empirical evidence is needed to show how this can be realised.

Energy resilience is linked to broader community resilience. The International Federation of Red Cross and Red Crescent Societies defines community resilience as "the ability of individuals, communities, organisations or countries exposed to disasters, crises and underlying vulnerabilities to anticipate, prepare for, reduce the impact of, cope with and recover from the effects of shocks and stresses without compromising their long-term prospects" (IFRC, 2014). Norris et al. (2008) argue that all communities have the potential to function effectively and adapt successfully in the aftermath of disasters and that community resilience emerges from four primary sets of adaptive capacities that come together to form strategies for disaster readiness – economic development, social capital, information and communication and community competence.

Other sectors have also developed approaches that support communities in dealing with disasters. In the area of housing, communities are increasingly being recognised as active participants in the reconstruction process with the majority of affected families rebuilding their homes, using their own resources (Parrack et al., 2014). As a result, international agencies are providing support for this self-recovery process by providing material, financial and technical assistance to enable households to repair, build or rebuild their own shelters (Maynard et al., 2017). In the water, sanitation and hygiene (WASH) sector, work on resilience has tended to focus on adaption to climate change (Global Water Partnership and UNICEF, 2015; Oates et al., 2014). Pubic-private partnerships have been suggested as a way to engage business, government and communities (Johannessen et al., 2014).

Resilience can sometimes be used to further entrench old paradigms and 'vaccinate' communities so that they can take larger doses of inequality and environmental degradation, rather than addressing the root causes of global social inequality (Kaika, 2017). Initiatives aimed at building energy resilience must be explicit about what is being maintained in the name of resilience, for whom and by whom (Cretney, 2014). Equally, community and community-based approaches need critical reflection to ensure that communities are not treated as heterogeneous. This is true in energy-related work (Campbell et al., 2016) and as well as in disaster-related work (Titz et al., 2018).

Community energy resilience framework

Conceptual understanding of energy infrastructure as socio-technical systems is well developed and includes technological, institutional, financial, social and ecological aspects (Retnanestri et al., 2008). All aspects are interrelated and necessary for long-term energy service provision. This section discusses resilient energy infrastructure that is situated within broader community resilience.

Households and communities often use energy in a strategic way to improve their resilience in everyday life and in disaster situations (which is discussed further in the case study of Nepal). Acknowledging a community's own energy resilience strategies and including them in the planning process could lead to the development of more resilient energy infrastructure and more effective humanitarian assistance.

A framework for community energy resilience is shown in Figure 5.1. It shows that energy infrastructure is embedded in energy resilience, community resilience and governance structures in turn.

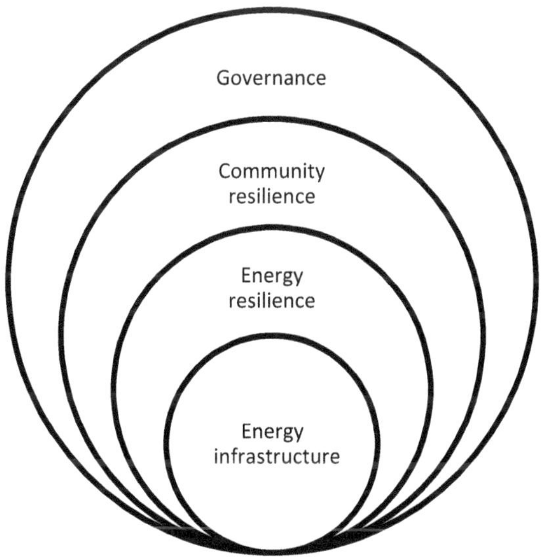

Figure 5.1 A conceptual framework for community energy resilience

Source: Author's own.

The framework in Figure 5.1 formed the basis for discussion at a series of workshops on the theme of 'Community energy resilience and electricity systems' which sought to gather international experience on this subject. Workshops have so far been held in the UK (2018) and Nepal (2019), with contributions from 38 experts from eight countries. The workshops focused on identifying research needs and opportunities to increase community energy resilience in on-grid, mini-grid and stand alone electricity systems.[1]

The first workshop in the UK confirmed the need for research at each of the boundaries of the framework in Figure 5.1. The workshop focused on technical innovations required to improve the resilience of electricity systems, the linkages between electricity systems and community resilience and the governance mechanisms needed to improve energy resilience at different scales. It also emphasised that energy resilience should focus on securing energy services and development outcomes, rather than security of energy supplies. The discussions also emphasised the need for inclusion and capacity development at all levels.

The second workshop, which took place in Nepal, focused on community energy resilience in the context of disaster risk reduction in the South Asia region. Discussions included the energy and resilience situation in South Asia, and identifying research needs in disaster response, recovery, prevention and preparation. On the technical side of energy infrastructure, quality control guidelines in systems design and installation as well as the integration of mini-grids into national grids were identified as potential areas for increasing the reliability of energy systems. However, the benefits to resilience need to be weighed against the possibility of increased costs to energy systems and implications for energy access. Thus, energy efficiency measures might play a vital role in improving resilience while achieving energy access targets. There is a need for a holistic approach, which encompasses operational, regional and national levels.

Examining the linkages between energy resilience and community resilience, current disaster responses in South Asia generally did not include energy specifically. More emphasis could increase opportunities to 'build back better' energy systems during reconstruction, including the ability to feed in to long-term development goals through productive uses of energy for agriculture and enterprises. Specific challenges faced by informal urban communities and communities that have been displaced were discussed. Community organisation and social capital were recognised for the critical role that they play in restoring energy services.

In terms of governance, there is a need to balance the changing nature of exposure to risks, natural and built systems, users and managers of energy and rules and regulations. The energy security context of many national discussions around energy in the South Asia region was also highlighted.

Greater understanding of community energy needs and resilience strategies could improve the design of energy solutions. The following case study of Nepal highlights the importance of communities in coping after disasters.

Case study of Nepal after the 2015 earthquake

Nepal, a mountainous country in South Asia, has great renewable energy potential, including hydro power and solar energy. Nepal has succeeded in harnessing about 1,100 MW of hydroelectricity and about 50 MW of solar electricity. While investment in larger infrastructure has remained a challenge, mountainous terrain has made it difficult to extend the national grid to rural areas. Currently, about 76% of the population has access to national grid and about 12% have access to off-grid renewable energy solutions (ADB, 2017).

Nepal lies on the fault line of the Indian and Eurasian tectonic plates and is the eleventh most earthquake-prone country in the world (NPC, 2015). While minor earthquakes are common, Nepal has only experienced a major earthquake every few generations, with the last great earthquake in 1934. However, in April and May 2015, two major earthquakes hit central Nepal, along with over 300 aftershocks. These earthquakes took the lives of nearly 9,000 people, injured more than 22,000 people and destroyed over a half a million homes (NPC, 2015). The earthquakes forced 2.6 million people to flee their homes and was the largest single displacement event of 2015 associated with a disaster (Bilak et al., 2016). Most of the displaced people initially stayed in makeshift shelters near their homes, but many of the poorest households in the worst-affected rural areas sought shelter in collective urban sites in Kathmandu (Bilak et al., 2016).

An assessment by the National Planning Commission (NPC) of Nepal has an estimate of over USD 7 billion of damage caused by the earthquakes. Of that figure, USD 212.42 million was the estimated damage in the electricity sector (NPC, 2015). Figure 5.2 shows that infrastructure accounted for 9% of the damage

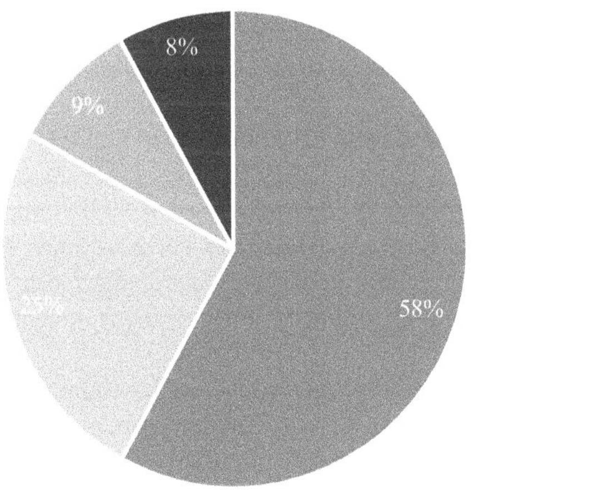

Figure 5.2 Share of damage by sector

Source: NPC, 2015.

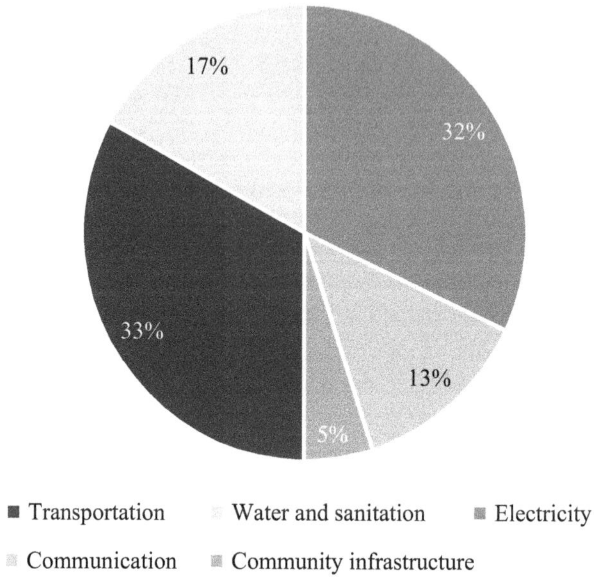

Figure 5.3 Share of damage in the infrastructure sector

Source: NPC, 2015.

caused by the earthquake. Figure 5.3 shows that one-third of that damage was to the electricity infrastructure.

While these figures reflect the damage to the grid-connected electricity infrastructure managed by the National Electricity Authority, the damage to standalone household energy systems was not included in this assessment. In the 14 districts most badly affected by the earthquake, the Government of Nepal's Alternative Energy Promotion Centre (AEPC) reported that there were a total of 145 micro-hydro systems providing electricity to over 37,000 households, as well as over 80,000 households accessing electricity via solar home systems and accessing electricity and over 79,000 households accessing clean cooking via biogas digesters. The APEC estimated that 30–50% of these systems were damaged by the earthquakes.

Though the earthquake has severely shaken the country, the people of Nepal have found ways to rebuild their lives. The Government of Nepal has focused on 'Building Back Better'. For the reconstruction of the private houses, the government has provided standard engineering designs and technical support through hundreds of engineers working with municipal governments through the newly established Nepal Reconstruction Authority. One of the initiatives to improve the resilience of these homes is the incorporation of solar home systems and household biogas (where applicable) in the standard designs. Through the AEPC, rehabilitation of damaged micro-hydro projects and biogas plants have been carried out in earthquake-affected districts, but this has not been completed in all districts due to lack of resources.

While the Government of Nepal is supporting households in the reconstruction process, people have also been putting in their own effort with locally available resources. A pilot study conducted in Nepal after the 2015 earthquakes indicated that communities leveraged local technical expertise, resources and informal networks to access energy services using renewable energy (To et al., 2016). The study showed that households were able to make repairs to some parts of their renewable energy systems. These included reassembling of solar home systems by collecting components from the damaged houses, extending the biogas pipeline to use in temporary shelters, partial use of micro-hydro facilities by disconnecting the supplies in the damaged houses and quick construction of improved mud cook stoves in temporary shelters. Other kinds of damage required external engineering expertise and were more difficult to organise, such as repairs to the dome of biogas plants and damage to micro-hydro equipment. Many households also switched cooking fuels to firewood as it was the most widely available and accessible option after the earthquake and during the subsequent blockade of the Indian border (Herington and Malakar, 2016).

Local organisations were critical in restoring energy access in the immediate aftermath of the earthquake. Community Rural Electricity Entities (CREEs) are local organizations that buy electricity in bulk from the national utility (Nepal Electricity Authority) and managed local distribution in areas previously not serviced by the grid. Most of the distribution infrastructure serviced by CREEs were damaged in the earthquake and local CREE technicians were able to restore supply to some households using temporary means. However, they did not have the required tools and procedures in place to do so safely (Buraityte et al., 2016). The Nepal Electricity Authority was hindered in its ability to respond quickly by its weaker presence in rural areas, difficult accessibility after the disaster and the need for repairs of the distribution system over a large geographical area. At the same time, longer bureaucratic processes and the need for large investments to repair the damage added to the difficulties for the National Electricity Authority in restoring the national grid in all areas quickly.

Informal networks were an important way for communities to support each other and maintain access to services after the earthquake in Nepal (Carrero et al., 2019). Many experts were able to draw on local knowledge about renewable energy during the disaster because of the large investment in capacity building at the local level for the scale up of renewable energy deployment in rural areas (Clemens et al., 2010).

In this case, several elements of Norris et al's. (2008) concept of community resilience have come to the fore. Communities relied on existing social capital, information and competencies to deal with disruptions to their energy services after the earthquake. This was especially important in the immediate recovery phase but also in the reconstruction. However, some types of damage were beyond the ability of communities and local organisation to repair, and there is a need to explore how they can be better supported in the future, including training on restoring energy services safely, insurance mechanisms and standing contracts or suppliers lists for emergencies.

This case study has shown that where the public awareness and local technicians are available, informal networks can be more effective than the formal networks in hard to reach locations. Government bodies and international organisations need to coordinate with and support informal networks and local organisations in reinstating energy access post-disaster.

Discussion and conclusions

The concept of community energy resilience intersects with some key developments within humanitarianism and forced migration. Classical humanitarianism focuses on international actors such as the United Nations, international donors and international nongovernment organisations. The emerging concept of 'Resilience humanitarianism' sees humanitarian governance as "emerging through interactions between authorities, implementing agencies and communities" (Hilhorst et al., 2019).

Displacement hits hard at the heart of resilience as displaced communities are less resilient. There is evidence to show that internally displaced persons (IDP) have lower levels of resilience due to lower levels of social capital when compared with other local communities. Community networks and trust, religious intuitions and self-help organisations as well as interactions with other ethnic communities, government and nongovernment organisations are all disrupted for IDP communities (Jayawardana et al., 2019)

The World Humanitarian Summit in 2016 committed to bridging short-term humanitarian and longer-term development efforts to improve coherence in the international system. This is especially true in areas experiencing ongoing disruption due to protracted conflict and climate change. The Summit emphasised the need to reinforce the coping strategies of communities and prioritise the development of local and national capacity to respond to crises (Ban, 2016). In the energy sector, this will require strengthening the links between community resilience and energy resilience.

This chapter has begun to explore the research needed on energy resilience and the community responses to energy challenges in terms of preventing, preparing, responding and recovering from disasters. While it was a great tragedy for Nepal, the 2015 earthquake highlighted what communities and local organisations can do to restore and access energy after disasters where huge numbers of people were internally displaced. Further research on identifying community energy resilience strategies, systematic measurements for community energy resilience, as well as improved energy infrastructure design and planning could improve outcomes in the future. Energy interventions should work to enhance existing community energy resilience strategies. In this way, we can more effectively ensure access to energy services in the face of uncertainty and disruption.

Note

1 The workshops described were part of the 'Research collaborations for community energy resilience in low-income countries' project, a collaboration between the Energy and Economic Growth Applied Research Programme and the UK Low Carbon Energy for Development Network, with funding from the UK Energy Research Centre Whole Systems Networking Fund. The project was led by Long Seng To (Loughborough University), Simon Trace (Oxford Policy Management) and Ed Brown (Loughborough University).

Bibliography

ADB (2017), 'Nepal Energy Sector Assessment, Strategy and Road Map', *Asian Development Bank*, Manila, Philippines.

Adger, W. N. (2000), 'Social and Ecological Resilience: Are They Related?' *Progress in Human Geography*, 24 (3), 347–364. https://doi.org/10.1191/030913200701540465.

Ban, Ki-moon. (2016), 'World Humanitarian Summit Chair's Summary: Standing up for Humanity: Committing to Action', *United Nations*, Istanbul. www.agendaforhumanity.org/sites/default/files/resources/2017/Jul/Chair%27s_Summary.pdf.

Bazilian, M. and Chattopadhyay, D. (2016), 'Considering Power System Planning in Fragile and Conflict States', *Energy for Sustainable Development*, 32 (June), 110–120. https://doi.org/10.1016/j.esd.2016.02.011.

Berkes, F., Colding, J. and Folke, C. (2008), 'Navigating Social-Ecological Systems: Building Resilience for Complexity and Change', *Cambridge University Press*, Cambridge.

Bilak, A., Cardona-Fox, G., Ginnetti, J., Rushing, E. J., Scherer, I., Swain, M., Walicki, N. and Yonetani, M. (2016), 'Global Report on Internal Displacement 2016'. *International Displacement Monitoring Centre*. www.internal-displacement.org/globalreport2016/pdf/2016-global-report-internal-displacement-IDMC.pdf.

Buraityte, A., Maharjan, R., Sharma, M. and Ghimire, D. (2016), 'Disaster Risk Reduction Capacity of Community Rural Electricity Entities', Briefing paper 1. Resilience Initiative in Mechi-Mahakali-Mustang-Marchawar (RIM4). Discussion Series. Kathmandu, *Institute of Social and Environmental Transition-Nepal*, Nepal.

Campbell, B., Cloke, J. and Brown, E. (2016), 'Communities of Energy', *Economic Anthropology*, 3 (1), 133–144. https://doi.org/10.1002/sea2.12050.

Carrero, R., Acuto, M., Tzachor, A., Subedi, N., Campbell, B. and Long Seng To. (2019), 'Tacit Networks, Crucial Care: Informal Networks and Disaster Response in Nepal's 2015 Gorkha Earthquake', *Urban Studies*, 56 (3), 561–577. https://doi.org/10.1177/0042098018810606.

Cherp, A. and Jewell, J. (2014), 'The Concept of Energy Security: Beyond the Four As', *Energy Policy*, 75 (December), 415–421. https://doi.org/10.1016/j.enpol.2014.09.005.

Clemens, E., Rijal, K. and Takada, T. (2010), 'Capacity Development for Scaling Up Decentralized Energy Access Programmes: Lessons from Nepal on Its Role, Costs, and Financing', *United Nations Development Programme and the Alternative Energy Promotion Centre*, Rugby.

Cretney, R. (2014), 'Resilience for Whom? Emerging Critical Geographies of Socio-Ecological Resilience', *Geography Compass*, 8 (9), 627–640. https://doi.org/10.1111/gec3.12154.

Francis, R. and Bekera, B. (2014), 'A Metric and Frameworks for Resilience Analysis of Engineered and Infrastructure Systems', *Reliability Engineering & System Safety*, 121 (January), 90–103. https://doi.org/10.1016/j.ress.2013.07.004.

Gallopín, G. C. (2006), 'Linkages between Vulnerability, Resilience, and Adaptive Capacity', *Global Environmental Change, Resilience, Vulnerability, and Adaptation: A Cross-Cutting Theme of the International Human Dimensions Programme on Global Environmental Change*, 16 (3), 293–303. https://doi.org/10.1016/j.gloenvcha.2006.02.004.

Global Water Partnership, and UNICEF (2015), 'Integrating Climate Resilience into National WASH Strategies and Plans', *GWP & UNICEF*, Stockholm and New York.

Grove, K. (2018), *Resilience*, Key Ideas in Geography, Routledge, Taylor & Francis Group.

Herington, M. J. and Malakar, Y. (2016), 'Who Is Energy Poor? Revisiting Energy (in) Security in the Case of Nepal', *Energy Research & Social Science*, 21 (November), 49–53. https://doi.org/10.1016/j.erss.2016.06.025.

Hilhorst, D., Desportes, I. and de Milliano, C. W. J. (2019), 'Humanitarian Governance and Resilience Building: Ethiopia in Comparative Perspective', *Disasters*, 43 (S2), S109–S131. https://doi.org/10.1111/disa.12332.

Holling, C. S. (1973), 'Resilience and Stability of Ecological Systems', *Annual Review of Ecology and Systematics*, 4 (1), 1–23. https://doi.org/10.1146/annurev.es.04.110173.000245.

IFRC (2014), 'IFRC Framework for Community Resilience', *International Federation of Red Cross and Red Crescent Societies*, Geneva. http://preparecenter.org/sites/default/files/everyone_counts_en.pdf

IPCC (2014), 'Annex II: Glossary', Climate Change 2014: Synthesis Report Contribution of Working Groups I, II and III to the Fifth Assessment Report of the Intergovernmental Panel on Climate Change, 117–130. *IPCC*, Geneva, Switzerland.

Jayawardana, J., Priyantha, R., Magni, M. and Marincioni, F. (2019), 'Disaster Resilience among War-Affected People Resettled in Northern Sri Lanka: Challenges Revisited', *International Journal of Disaster Risk Reduction,* 34 (March), 356–362. https://doi.org/10.1016/j.ijdrr.2018.12.005.

Johannessen, Å., Rosemarin, A., Thomalla, F., Swartling, A. G., Stenström, T. A. and Vulturius, G. (2014), 'Strategies for Building Resilience to Hazards in Water, Sanitation and Hygiene (WASH) Systems: The Role of Public Private Partnerships', *International Journal of Disaster Risk Reduction*, 10 (December), 102–115. https://doi.org/10.1016/j.ijdrr.2014.07.002.

Kaika, M. (2017), ' "Don't Call Me Resilient Again!": The New Urban Agenda as Immunology . . . or . . . What Happens When Communities Refuse to Be Vaccinated with "Smart Cities" and Indicators', *Environment and Urbanization,* 29 (1), 89–102. https://doi.org/10.1177/0956247816684763.

Kruyt, B., van Vuuren, D. P. de Vries, H. J. M. and Groenenberg, H. (2009), 'Indicators for Energy Security', *Energy Policy, China Energy Efficiency*, 37 (6), 2166–2181. https://doi.org/10.1016/j.enpol.2009.02.006.

Maynard, V., Parker, E. and Twigg, J. (2017), 'The Effectiveness and Efficiency of Interventions Supporting Shelter Self-Recovery Following Humanitarian Crises', *Oxfam, Feinstein International Center, UKAID*, Oxford. https://doi.org/10.21201/2017.8739.

Molyneaux, L., Brown, C., Wagner, L. and Foster, J. (2016), 'Measuring Resilience in Energy Systems: Insights from a Range of Disciplines', *Renewable and Sustainable Energy Reviews,* 59 (June), 1068–1079. https://doi.org/10.1016/j.rser.2016.01.063.

Norris, F. H., Stevens, S. P., Pfefferbaum, B. J., Wyche, K. F. and Pfefferbaum, R. L. (2008), 'Community Resilience as a Metaphor, Theory, Set of Capacities, and Strategy for Disaster Readiness', *American Journal of Community Psychology*, 41 (1–2), 127–150. http://dx.doi.org/10.1007/s10464-007-9156-6.

NPC (2015), 'Nepal Earthquake 2015 Post Disaster Needs Assessment', *National Planning Commission, Government of Nepal*, Kathmandu.

Oates, N., Ross, I., Calow, R., Carter, R. and Doczi, J. (2014), 'Adaptation to Climate Change in Water, Sanitation and Hygiene', *Overseas Development Institute*, London.

O'Brien, G. and Hope, A. (2010), 'Localism and Energy: Negotiating Approaches to Embedding Resilience in Energy Systems', *Energy Policy*, Special Section: Carbon Reduction at Community Scale, 38 (12), 7550–7558. https://doi.org/10.1016/j.enpol.2010.03.033.

Ouyang, M. (2014), 'Review on Modeling and Simulation of Interdependent Critical Infrastructure Systems', *Reliability Engineering & System Safety,* 121 (January), 43–60. https://doi.org/10.1016/j.ress.2013.06.040.

Panteli, M. and Mancarella, P. (2015), 'The Grid: Stronger, Bigger, Smarter?: Presenting a Conceptual Framework of Power System Resilience', *IEEE Power and Energy Magazine*, 13 (3), 58–66. https://doi.org/10.1109/MPE.2015.2397334.

Parrack, C., Flinn, B. and Passey, M. (2014), 'Getting the Message Across for Safer Self-Recovery in Post-Disaster Shelter', *Open House International,* 39 (3). https://radar.brookes.ac.uk/radar/items/b0bd437a-4aec-4849-a849-fb3f8fc2f0ce/1/.

Retnanestri, M., Outhred, H. and Healy, S. (2008), 'The I3A Framework – Enhancing the Sustainability of Off-Grid Photovoltaic Energy Service Delivery in Indonesia', *Jurnal Ilmiah Teknologi Energi*, 1 (7).

Roege, P. E., Collier, Z. A., Mancillas, J., McDonagh, J. A. and Linkov, I. (2014), 'Metrics for Energy Resilience', *Energy Policy*, 72 (September), 249–256. https://doi.org/10.1016/j.enpol.2014.04.012.

Sharifi, A. and Yamagata, Y. (2016), 'Principles and Criteria for Assessing Urban Energy Resilience: A Literature Review', *Renewable and Sustainable Energy Reviews*, 60 (July), 1654–1677. https://doi.org/10.1016/j.rser.2016.03.028

Spyrou, E., Hobbs, B. F., Bazilian, M. D. and Chattopadhyay, D. (2019), 'Planning Power Systems in Fragile and Conflict-Affected States', *Nature Energy*, 4 (4), 300. https://doi.org/10.1038/s41560-019-0346-x.

Titz, A., Cannon, T., Krüger, F., Titz, A., Cannon, T. and Krüger, F. (2018), 'Uncovering "Community": Challenging an Elusive Concept in Development and Disaster Related Work', *Societies*, 8 (3), 71. https://doi.org/10.3390/soc8030071.

To, L. S., Subedi, N., Campbell, B., Carrero, R. and Mulugetta, Y. (2016), 'Enhancing Community Resilience Using Renewable Energy in Nepal: Preliminary Results'. https://doi.org/10.6084/m9.figshare.5579380.v1.

6 Incentivizing market mechanisms for access to energy

Raffaella Bellanca

Market dynamics in humanitarian settings

Market systems and market approaches

There is a widespread perception in the humanitarian sector that market systems are not suitable to the conditions of relief settings. This is mainly based on the beliefs that displaced people do not have sufficient purchasing power, that immediate assistance is incompatible with markets and that the mandate of relief agencies should be limited to addressing basic needs.

This chapter will clarify how the expression "market system" is used in this context and will challenge these beliefs by arguing that in many humanitarian situations the necessary conditions are present for markets to deliver efficiently when considering energy services. Short-termism and the narrow focus on imminent survival without a longer-term vision, on the contrary, risk deepening the crises that interventions are meant to address and further, inhibiting recovery. First, immediate crisis situations have a sustained tendency to transform into protracted ones (Grafham and Lahn, 2018).[1] Second, in displacement settings people in many cases do pay for services, and often at higher prices than in other contexts. Furthermore, the ability to pay in full for goods and services may not be the real issue, as much as the delivery mechanism (i.e., the way in which the payment is carried out and the service is delivered). This is because not all clients must be able to afford the products, since the most vulnerable can be supported to purchase through aid cash and/or vouchers.[2] In addition, several other ingredients also make up a "market system", including established distribution networks, demand and supply dynamics, customer service and the imperative to adapt to market trends that, importantly, are made by customer's needs and wants rather than procurement choices imposed from outside. Market approaches have the potential to preserve existing local business dynamics, shape a market landscape of products that are relevant and appropriately priced for customers, reduce tensions with host communities and prevent a spillage of goods sold for cash at distorted prices.

Humanitarian agencies are increasingly reviewing their strategies adding to emergency response longer-term investments in social protection and safety nets, reducing risk, building resilience and providing beneficiaries with the tools to prevent disaster or recover faster from it. The World Food Program (WFP) for example adds to food distributions that reactively respond to famines, proactive work on strengthening Food Systems from production to transformation and consumption. According to WFP, delivering services through local markets by transferring cash into the local economy not only improves food and nutrition security but also provides a substantial boost to local economies (WFP, 2019). Cash transfers are both a short-term, direct mean of meeting basic needs and also have a proven multiplier effect on local economies and enhance self-reliance (Bastagli, 2016). For example, studies in Lebanon, Rwanda and Uganda have shown that USD1 given in cash to a refugee or vulnerable person translates into about USD2 in the local economy (WFP, 2019). According to ODI, injecting this capital, where feasible, has several advantages including: reducing the costs of delivering humanitarian aid and so make limited budgets go further; supporting local markets, jobs and incomes of local producers; increasing financial inclusion by linking people with payment systems; and most importantly, providing affected populations with choice, more control over their own lives and dignity (ODI, 2015).

At the individual and household level, humanitarian aid is intended to address the most immediate survival needs. However, it is through functioning markets that services and products are delivered in the longer run, hence the recovery of markets is pivotal to lifting communities out of crises faster. Cooking, lighting, communication, heating or cooling are the immediate necessities for individuals. These services are provided by entities such as businesses and governments within and outside the community, linked to yet larger systems of nation wide and international actors that import, transport, produce and process. Although these systems are complex, in many humanitarian situations (not all, but many), they are not fully disrupted or are recovering or in some cases are already functional and can efficiently react to delivering energy products and services. Strengthening those systems rather than building parallel delivery structures contributes to creating economic opportunities in the longer term.

And yet, many aid efforts overlook the social fabric of communities and the role that private and local actors play within it. Rather than leveraging their capacities and encouraging their recovery, agencies often bypass enterprises that exist in the host community and those that could emerge and flourish in the camps by providing goods and services themselves. A publication by ODI analyzing the behaviour of markets during relief interventions in Pakistan, Mali and South Sudan comes to the conclusion that

> In-kind aid replaced demand for goods in the market. As well as the direct impact on trade volumes, this also undermined the confidence within the market chain necessary for trust in credit to be restored. This constrained the recovery of market chains.
>
> (Levine, 2017)

A huge potential is lost in the process by ignoring existing formal and informal market structures; in the words of a publication by the University of Oxford,

"Refugee communities are often integrated within vibrant and complex economic systems. Recognising and understanding this represents an opportunity to turn humanitarian challenges into sustainable opportunities" (Betts et al., 2014).

Alternative mechanisms to distributions of goods such as cash and voucher transfers are normally applicable and have the advantages of forging crucial links between customer and supplier, avoiding market distortions, allowing end users to appreciate the real value of goods and letting them choose between different options to ensure that goods are relevant to their needs. Providing cash transfers acknowledges that every household and business has unique needs.

Beside cash transfers that focus on the demand side of value chains, there must be interventions aiming to support the supply side as well. An interesting approach is WFP's Pro-Smallholder Food Assistance that aims to build resilience and market access for smallholder farmers starting with sourcing locally the food for its school feeding programs, including in camps (WFP, 2017).

Purchasing power of displaced people

Life conditions for displaced people vary greatly depending on several factors, including their own socio cultural and economic status prior to the crisis, gender, age, ethnicity, geography of origin, whether they found refuge in a camp or in urban areas within host communities and crucially, how long they have been in the situation. It is by no means a homogeneous group of people, some having been able to preserve few assets while fleeing, others having started economic activities in the new location or receiving remittances from relatives.

Research conducted in Kakuma and Kalobeyei refugee settlements in northwest Kenya by Mastercard and Western Union has analyzed the diverse streams

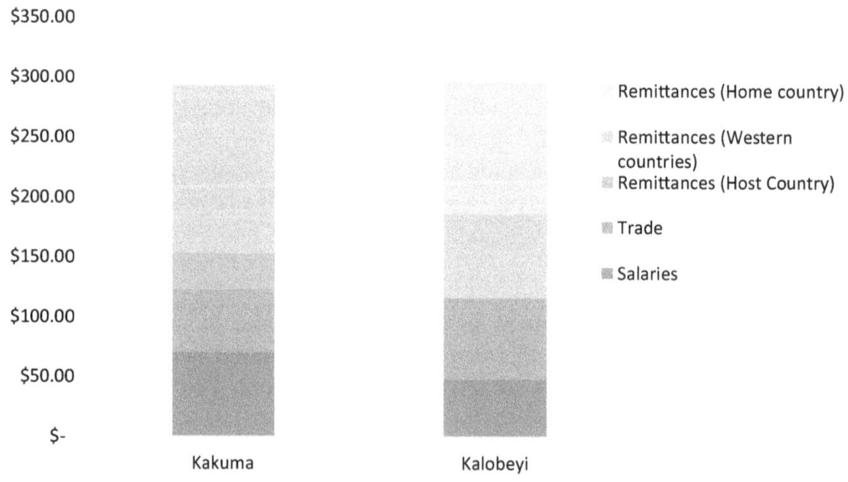

Figure 6.1 Streams of income for refugees in Kakuma and Kalobeyi (per month)

Source: Data from Mastercard and Western Union, 2017

of income of refugees. As shown in Figure 6.1, these comprise salaries from agencies, remittances and trade. In Kakuma, monthly salaries of people working for agencies amount to $70 in average, income from trade to $53, remittances from friends and family who have relocated from the camp to Western countries to $140 per household, or who have moved to the host country ($30). In Kalobeyei, a newer camp where social networks are yet to be established, the figures from salaries and trade are comparable ($48 and $68 respectively), but remittances only come from links back home ($110) or from friends and family in Western countries ($70) (Mastercard and Western Union, 2017).

Given the breadth of in-kind assistance in these camps, disposable income goes beyond food, covering also nonessential needs, such as education, health and transport. In Kalobeyei refugee camp, a site more isolated and lacking main services, all expenses are considerably higher than in Kakuma. This includes sending children to boarding school ($21 here vs $2 in Kakuma), reaching the hospital ($36 vs $9) and general transport ($38 vs $7) (Mastercard and Western Union, 2017).

The study identifies a market segmentation with three groups of refugees, characterized by ethnicity and time spent in the camps. The "impoverished new arrivals", mostly South Sudanese, who arrived with few resources, have a limited social network within the camp and only receive small amounts of remittances from their home country, which lacks banking infrastructure and is in an economic crisis. The "intermediate group" that has been in the camps for some time and have some networks and social support but are not actively running businesses or trade. The "economically active", dominated by the Somali community, receive remittances from emigrated friends and family and enjoy social support from a strong community within the camp and with economic activities of trade within and outside the camp.

Other research in Tanzania confirms that revenues in Nyarugusu camp, similarly to the MasterCard study in Kenya, come from salaries by agencies, remittances and trade (Rivoal and Haselip, 2017). In the same study, over a third of respondents declared receiving cash transfers from WFP at some point rather than in-kind food rations, which had the effect of improving food diversity and allowing choices according to cultural food preferences. To overcome the reluctance of refugees to declare their income, the questions in this study have included information about income proxies such as ownership of assets, employment or having an M-Pesa account.[3] Ten percent of the interviewees reported receiving remittances from family or friends living abroad (often refugees who had emigrated to Western countries, mostly US and Canada). About a quarter mentioned at least one family member engaged in formal employment from agencies within the camp (which by law of the Tanzanian government cannot exceed $26.70 monthly). Although the encampment policy limits the right to work, during the focus groups it became clear that most refugees are involved in some form of income generating activity inside or outside the camp (Rivoal and Haselip, 2017).

Looking at proxies, almost half (44%) of the interviewees reported using mobile banking, indicating the existence of a cash-economy. In addition, 43%

of the households are using solar PV panels for which there has been no external distribution or subsidy in the camp, revealing both an ability and willingness to pay for modern and clean, energy services.

Food and fuel are the main expenses incurred by refugees, reaching in some cases 50% of average wages. The study reports that most refugees collect firewood, but the remaining buy firewood or charcoal, and 23% do both. People who buy their fuel, as opposed to collecting or bartering food rations for it, are better off: 78% of the households receiving money from abroad buy their fuel and M-Pesa users, owners of solar panels and families with working members, have higher fuel expenditures.

Research by the MEI about spending patterns of refugees in Burkina Faso (Goudoubo camp) and Kenya (Kakuma and Dadaab) finds that high percentages of interviewees incur monthly expenditures – reaching over $70 in Goudoubo and Dadaab – of which energy accounts for about 15% to 20% on average (Vianello, 2016; Okello, 2016; Corbyn and Vianello, 2018), but for some families it costs up to 50% of their income. As an example, in Kakuma and Goudoubo, families' monthly costs for cooking fuel, without considering other energy needs, amounts to USD5 and USD6 respectively, from a total average income of USD42 and USD76 respectively (Corbyn and Vianello, 2018). Meanwhile, in Goudoubo, households who use low-quality torches spend six times more on light than those that own a quality solar product.

Although this field of research is still limited and more efforts would be needed to provide a clearer picture over the entire refugee population and within sub-populations, including gender and age differences, it already highlights that market dynamics and economic activities are present in camps. It also shows that economic activities are intrinsically linked to the host community, that there is a clear market segmentation between and within camps and that refugees are already paying, and dearly, for energy services.

Energy delivery models (EDM)

We have so far proposed that it would be beneficial to adopt market approaches in the humanitarian sector to meet energy needs. However, the definition of what a "market approach" is, and what it means to adopt it, still lacks a clear definition.

Adopting a market perspective requires shifting from a technology, procurement and distribution-centred focus, to considering the design of energy delivery models (Bellanca et al., 2013a; Bellanca, 2013b; Bellanca, 2014) in their entirety, including consideration around the value chain, socio cultural context and enabling environment. It means assessing the needs of displaced persons in a holistic way, considering both households and businesses, mapping the existing energy infrastructure and services and investigating the local regulatory landscape. This type of analysis is commonly carried out to assess food, shelter and other needs at times of crises,[4] and there is a stark need to apply it to energy too. In assessing local and international competitive markets, preexisting links can be leveraged with private sector actors and distribution networks ensuring the availability of energy goods through the most efficient distribution method available. It is this comprehensive design that underpins the solution to the major energy challenges faced by the humanitarian sector.

Depending on the situation, interventions may be more or less market-reliant. "Light touch" energy delivery models can entail supporting private sector engagement, in particular local businesses, in meeting the energy needs of households and communities and would likely include cash-based interventions to support demand. Fully competitive market solutions might entail the outsourcing of energy services altogether, for example for power generation or supply of energy consuming goods.

What underlies all the different approaches, however, is the constant focus on the design process with the commitment to propose solutions that lay the foundations for longer-term sustainability without impacting on promptness in the response.

Fifty shades of market

Distributions should only take place in the absence of other solutions; for example, when commercial distribution channels have been disrupted in the aftermath of the crisis or have never been present and goods are not available. Even in these cases, the first step is to identify the barriers that are preventing the private sector from operating and adopt distribution approaches that bring products where they are needed in the immediacy but also build more permanent penetration channels. Access to energy is an issue that bridges from the humanitarian to the development contexts and from refugees to host communities. Any initial action will lay the foundations for a system that can either be conducive, stimulate recovery and long-time prosperity or may even hinder it.

In 2017, Hurricane Maria destroyed much of Puerto Rico's already aging electricity infrastructure. An assessment from the international organization Mercy Corps revealed an underdeveloped solar PV sector in the island across the entire spectrum of solar products, from lanterns to solar home systems and mini-grids. For example, in some areas, the only solar product stocked by the shop chain Home Depot (garden solar lights) completely sold out after the hurricane. The organization's acute response consisted of cash transfers and the distribution of solar lamps with charging capabilities and water filters. These were procured through the local network of shops of oil and gas company Total, operating in the country and serving the local population with 210 fuel stations since 2004.

Even distributions can therefore be used sensibly and ingeniously to hopefully lead to more sustainable practices (market light), however, interventions that use funds to create long lasting changes by supporting or establishing value chains in the system (market heavy) have a more powerful impact.

Market models for cookstoves

The cookstoves sector, where the challenges encountered in meeting the needs of the millions that still prepare food by burning woodfuel in and around refugee camps, offers a particular case that has so far proven difficult to address.

In too many cases, activities initiated by humanitarian agencies consist of stove and/or fuel distributions to beneficiaries. Even when high-quality stoves are distributed, without other supporting actions their adoption rates can remain low. In some cases, local communities are encouraged and trained to build tier 1 mud stoves, in the absence of proper quality control mechanisms, by relief organizations which then procure these stoves, store them in warehouses and distribute them to beneficiaries. This despite the fact that a range of artisanal stoves, of similar and often better quality, are often present and already sold in local markets.

The funds dedicated to core functions such as procuring stoves (on behalf of the end users) and distributing them (on behalf of markets), could be used instead to stimulate demand and encourage supply, letting private sector channels deliver the service and the end users choose the products. This would also allow for market segmentation, diversification of products and meeting the needs of those who are able to pay and already do so, as well as designing leaner subsidies for the ones who do not. Ideally, subsidies would be given in the form of mobile money or vouchers as part of an overall social protection system, and that would allow the recording of all digital payment transactions in a central database. Collecting this data would also enable agencies to better understand what really matters to beneficiaries, and which trade-offs between quality, price and other parameters, they are willing to take.

In Bangladesh a collaboration between FAO, WFP, UNICEF and IOM is providing refugees identified in WFP's beneficiary registration system, with cards that are used to access food and hygiene items from WFP's selected retailers. Similarly, LPG stoves, canisters and refills have been added to the portfolio through an agreement with Total and their retailers.

Lessons from longer-term development programming show that building certified efficient stoves of consistent quality locally entails significant upfront investments and an entrepreneurial mindset. This is needed to successfully and sustainably create or strengthen entire value chains starting from semi-industrial production facilities all the way to last mile distribution. The introduction of alternative fuels to firewood and charcoal, such as Liquid Petroleum Gas (LPG), ethanol or briquettes and pellets from agricultural residuals, presents similar challenges. In other words, creating entire value chains from scratch is neither cheap nor easy, and this often pushes organizations to choose less costly but ineffective shortcuts.

The economic argument can also be challenged since when the whole cooking system, made of stove and fuel, is considered, investments over the medium to long term become evidently more favourable to solutions that are more efficient. Using less firewood, or eliminating the need for it, decreases or avoids environmental and social challenges such as fuel linked deforestation and exposure to risk for fuel collectors.

To build market systems can be complex and expensive for relief agencies but building from the ground up is not the only possibility. Where the economic opportunity is clear, the private sector can be attracted to co-develop and invest in new markets including humanitarian settings, helping to bypass both the investment and set-up barriers.

Key to scoping the market opportunity and stimulating this engagement, is to undertake an effective market analysis. This analysis can estimate the market size, customer purchasing power, aspirations and preferences, identify existing market players, design robust business models, compare and cost competing offers and finally, ensure the buy in from other key stakeholders such as governments and host communities. An onerous preparation work that is often neglected precluding a priori the creation of effective market dynamics.

Another aspect that characterizes humanitarian interventions and does not resonate well with competitive markets, is the tendency to propose homogenous solutions across the board. Recognizing the diversity in needs, and willingness to pay, among and within refugee communities, it is important to expand the offer of cooking appliances beyond the very basic models to include aspirational products targeted at a diversified customer base. This is constituted not only by the poorest households, often the target of humanitarian organizations, but importantly, also made of small businesses, better off "early adopters" and figures of influence in the community. Encouraging commercial and institutional players, who already pay for their fuel, to switch to cleaner ones is arguably feasible (Rivoal and Haselip, 2017) and highly desirable, for many social and particularly environmental reasons given the environmental impact that their higher consumption level imply. LPG, ethanol and briquettes are all very attractive propositions for this market, particularly associated with PAYG payment models and an increasing number of interventions are now looking at the opportunities for non-wood solutions in humanitarian settings (Rivoal and Haselip, 2017).

Showcasing clean cooking practices and technologies in canteens catering for the UN organizations, street kiosks and restaurants has multiple advantages of setting the example, opening a market, raising awareness in the community at large and can also provide the ideal ground for testing the usability and appropriateness of different stove models in different camps.

Together with evidence of demand, experience from the Energizing Development programme[5] shows that an effective instrument to ensure supply by motivating manufacturers to establish or increase their presence in and around the settlement, strengthening their network of sales agents and retailers, is to provide economic incentives such as results based financing (Endev, 2018). Engaging private actors also has the additional advantage of creating a direct relationship between customer and vendor and helping to create new employment opportunities and a market-like environment that meets demand flexibly and guarantees maintenance and repairs.

Market models for fuels

In the humanitarian context as well as in development, practitioners often seem all too keen to design programs that take for granted a community cohesion and willingness to participate in activities on a voluntary basis that may or may not be present. In fact, in many displacement situations not only do people not know each other before arrival, but they may belong to completely different ethnic groups also speaking different languages. Social ties may require time to develop and sometimes never materialize due to deep cultural differences. Given the temporary settings, a sense of ownership and responsibility for the natural environment may not be present. In these cases, models that entail beneficiaries' collaboration and voluntary efforts for the common good, may be unrealistic, particularly when the benefits require years to materialize. And yet many reforestation or sanitation/energy solutions are still designed on these faulty assumptions.

To reach sustainable woodfuel use, increasing supply must be part of the equation. However, to ensure take up of forestry/reforestation schemes, interventions have to present clear and immediate benefits for the end users. Classic approaches to reforestation/ rehabilitation range from distributing seedlings, organizing awareness/planting campaigns and building tree nurseries. These interventions focus on short-term activities, while longer-term results are left to the voluntarily participation of beneficiaries, often failing to achieve impact when it does not occur. Rather than relying on the beneficiaries' willingness to contribute, interventions with aspirations to long lasting success could be based on business models that make engagement worthwhile for refugees: ascribing a value to the trees, describing the way in which they can be exploited and prescribing economic incentives that can motivate beneficiaries to invest time and effort to make them grow.

Models that look at cost analysis, considerations around key legal, social and environmental issues, stakeholder partnerships and financing mechanisms, are key to create interest in growing biomass, either purely as a sustainable source of energy or with other additional economic returns (e.g., fruits, fodder). Revenues from intercropping, agroforestry and carbon sequestration can contribute to both the upfront and long-term costs of sustainable reforestation/rehabilitation. For example, payment for environmental schemes (PES) and carbon accreditation initiatives, such as REDD+,[6] can generate income for the medium and long term, ensuring continuity of coordination and monitoring activities after the project's end-date.

During the time-lag, before trees begin to reach maturity, reforesting efforts require strong commitment. It is hard to imagine refugees who struggle to cover their basic needs, to be motivated forest stewards during this phase without tangible incentives. A retribution scheme for tree-caretakers, each responsible for a plot of land, could be, for example, based on remote satellite technology which would allow a project manager to measure accurately and relatively inexpensively biomass growth on the assigned parcels.

Other clear examples of applications where designing successful programs cannot ignore the importance of outlining the energy delivery model (including costs and revenues, ownership, management, maintenance, defining rights and responsibilities and all the key elements to run operations efficiently and profitably) come from circular economy concepts at the crossing between sanitation or waste disposal and energy. The potential for these systems to solve existing problems while creating much-needed resources such as energy, water and nutrients recovery is enormous. This is examined in more detail by Hakspiel and Lehne in Chapter 11.

Market models for solar products

Similar considerations around demand, supply and building value chains equally apply to any other technology (typically solar), energy need (lighting, cooling, heating, processing, device charging) or context (from isolated camps to conflict-prone areas and urban centres where already overloaded infrastructures can be further stretched by the influx of migrants).

The value of solar systems over diesel generators or unreliable grids has been demonstrated in most places but it is all the more evident for fragile, or conflict affected, contexts. These include refugee camps, of which South Sudan or Northern Kenya are examples but also regions stricken by conflict such as Afghanistan or Somalia. Here transporting fossil fuel is more expensive and smaller off-grid power generation is more resilient to disruption by armed groups. Common barriers to energy access across such locations are underdeveloped supply chains, poor economies of scale leading to high marginal costs for products and services, as well as nascent knowledge and skills existing within the local industry. Pioneering projects are needed to build confidence despite the volatile security situation and initiate a business-friendly environment that can attract commercial investors. While a donation for a single system is a drop in the ocean, using those funds to support the reestablishment of conducive market dynamics can bring scalable change.

Access to electricity for residences and businesses is very limited in Afghanistan, especially in rural districts with per capita power consumption among the lowest in the world. However, investment in power supply is listed as one of the top sectors for economic growth by the Afghan government. Despite the opportunity, the solar off-grid sector in the country is severely underdeveloped and uncompetitive leaving a gap that affects end users and their ability to access cheaper and more reliable energy services.

In a project that started in 2015, Mercy Corps used donor funded patient capital investment to prove the viability of a Sharia-compliant "Build Operate Lease Transfer of Ownership" business model that provides solar systems to Afghan customers. Mercy Corps acted as intermediary, providing technical expertise and financing for a 60KW off-grid solar PV hybrid system to the Bost University & Hospital (BUH) building, a privately run facility located in Laskhar Gah City provincial capital of Helmand province. The project showed that the unmet demand for power in the country can be sustainably addressed through solar PV installations on a full cost recovery basis. BUH repay time-lag for the loan is 36 months, while the expected lifetime of the system is 17 years.

Box 7.1 A personal story of the transition to solar power in Afghanistan

The building used to receive electricity from two generators that operated between 8:00 am to 6:00 pm and were affected by frequent malfunctions that required technicians to maintain. The installation of the solar equipment brought several benefits well beyond the economic gain. It has eliminated the generator noise and air pollution, improving the quality of care of the health facility. Greater reliability of electricity allows the medical staff constant access to power and longer service hours allow students to access the library furnished with digital textbooks and computers until 10:00 pm and the laboratory for educational purposes and medical tests.

> I remember a doctor shouting in the corridor, the scalpel in his hand, that he needed electricity. The generator, located in the basement, had stopped working. The doctor had to grab a torch and run back to the surgery room to continue to operate.
>
> Interview with Mr. Kazim, 10 January 2016, Mercy Corps

In the urban context of Jordan, Mercy Corps partnered with local private sector actors to address increased water and energy demand in host communities where displaced people from Syria found hospitality.

Aiming at supporting the creation of a market for efficient and renewable based energy products, the program in Jordan included financing revolving loans for solar water heaters and water-efficient technologies for households (aerators on faucets, showerheads, led bulbs). In 2012, the scheme was extended to mitigate the destabilizing effects on the water and sanitation sectors caused by the influx of Syrian refugees fleeing the civil war in Syria. When the program started, despite the abundance of solar energy, Jordan was affected by water shortages and faced serious challenges meeting its energy needs. Barriers included lack of awareness, distrust in technology caused by an influx of low quality products, high marketing costs for enterprises in remote areas and lack of finance to overcome the upfront costs for end users. Mercy Corps worked with private suppliers to build demand communities in northern Jordan through social marketing campaigns, capacity building of community-based organizations on efficient products, conservative behaviours and access to financing solutions for end users.

Outsourcing

At the end of the market engagement spectrum are solutions that move the implementation responsibility entirely away from humanitarians to specialized companies.

Although much of the focus for energy solutions in emergency situations is around the energy needs of beneficiaries, the biggest power consumers are actually humanitarian agencies themselves. Powering infrastructure in relief settings, where clinics, offices, communications towers and other necessary services have commonly been fuelled by diesel generators, represents a substantial economic burden for the organizations providing assistance.

This is where outsourcing the provision of energy is perhaps most applicable and indeed where the grid exists, agencies already purchase power from utilities, something that could be done in off-grid situations as well.

Recognizing that energy supply is only one side of the equation when addressing an organization's powering needs in humanitarian settings, and that moderating energy demand is crucial to overcoming perceived barriers about the cost of installing solar PV technology, it is important to keep considerations around energy economy and efficiency at the centre of the design process. For humanitarian operations this may include trialling direct current[7] systems and appliances and other low-energy consuming technologies, developing efficient shelters, planning landscapes that help to save energy with shadow or natural ventilation, introducing metering and automatic switch-offs together with behavioural change training packages for employees.

WFP has developed an ambitious Energy Efficiency Strategy and zero carbon emissions policy that is delivered by measuring, reducing and offsetting greenhouse gas emissions from operations. To this end, the Green Kit allows all WFP premises to assess and optimize their energy consumption with technical support from the engineering team; the Energy Efficiency Grants Programme offers partial funding to country offices and regional bureaux for energy-saving actions that demonstrate cost-effective reductions in emissions. In 2017, funding was awarded to six new projects, spanning five countries in Africa and Asia. Projects include lighting installations, high efficiency A/C units, solar water heaters and solar power solutions. More projects have been approved in 2018/2019. It is estimated that every USD 1 million invested in energy efficiency upgrades should result in operational cost savings of approximately USD 3 million over the next ten years, a threefold return.

Data is hard to find, but research from the MEI in the Burkina Faso and Kenya refugee camps, suggests that fuel spending (including for associated costs like maintenance and transportation) accounts for roughly 5% of humanitarian budgets (Grafham and Lahn, 2018). Fuel prices are volatile and not expected to decrease in the medium to long term. In addition, the cost of transporting fuel to often remote settlements is generally not considered and in South Sudan accounts for around 25%. Nor are internal or external running costs, bound to be significant for machines that require skilled maintenance and have poor reliability. Another issue is the decommissioning of purchased power generation systems that are often discharged well before the end of their expected lifetime due to lack of repair parts or funds for up-keep. Equipment can often lie abandoned at the end of a project, littering camps with no agency taking responsibility for recycling or dismantling.

This is unfortunate considering that settlements could represent the ideal situation to implement clean, reliable, economically attractive and more sustainable energy solutions such as renewable and hybrid systems that can generate benefits for displaced people, humanitarian agencies, international donors and the environment. They can also act as demonstration sites and innovation hubs to help catalyze off-grid energy solutions for host countries.

Humanitarian operations can seize this opportunity and excel in achieving SDG 7. However, despite the well-documented economic, social and environmental benefits, the application of renewable energy in the humanitarian sector is hampered by the upfront cost of solar and other renewable installations compared with diesel generators.

Should costs be calculated over the entire life cycle of the equipment and considering both initial investment but also operation, including the purchase of fuel, a very different picture would emerge. The expense of fuelling diesel generators can easily exceed the purchase price within three to five years, also because they often run inefficiently using 10–30% more fuel than needed. Moreover, a generator is expected to last for ten years (in practice less due to poor maintenance) compared to 20 years for a solar system (Arctic Energy Alliance, 2011). There is also a perception that camps are temporary structures and therefore not worth investing in (despite the estimation of 18 years for average camp lifetime) (Grafham and Lahn, 2018).

A way for agencies to bypass these concerns is to shift the paradigm from purchasing and operating equipment to paying for delivered services. This has the potential to revolutionize humanitarian operations by allowing the private sector to produce power, ensuring high technical standards together with cleaner, up to date and cheaper solutions, leaving agencies to what they do best, i.e., assisting beneficiaries. It is by no means a new concept, as agencies are routinely buying power from utilities where the grid is present, but it is still unusual in off-grid situations and seen with suspicion.

In this delivery model a technical partner finances, designs, procures, installs, operates and maintains the system; the receiving humanitarian agency signs a purchasing agreement for a time period between five to eight years (with current prices for equipment 2018) to allow the recovery of capital costs, interests, running expenses and creation of profit for the provider; a de-risking financing mechanism covering for unexpected early decommissioning of the camp is initially provided by a donor with the aim to become a mainstream insurance product once the concept is proven. The details of the economic model, such as exact ownership structures, minimum duration of the contract and monthly fees are flexible allowing for adjustments depending on circumstances. Carbon finance revenues or alternative financing mechanisms such as the "Peace Renewable Energy Credits" could potentially add to the concept attractiveness (Cook et al., 2019).

For this model to look competitive compared with inhouse power generation, it is important for humanitarian agencies to adequately track the costs of keeping a system running (fuel, servicing, replacing batteries, staff salaries) and consider those costs when drawing up energy budgets for relief operations.

The presence of an energy service provider stationed in the vicinity of a humanitarian settlement can also open entrepreneurial opportunities for displaced people and host communities. One of the biggest challenges for market-based solutions to energy access for refugees or, more generally, remote underserved communities, is the low energy consumption levels of potential customers. Without an anchor client that consumes and pays for commercially significant amounts of energy, suppliers are not able to run operations sustainably. However, once the solar system is established to serve humanitarian agencies, suppliers may be interested to increase the output and provide power on a commercial basis for businesses run by beneficiaries and host communities that are currently incurring high costs for fuel by using small diesel generators. This could simply mean adding a charging station, for phones and appliances, to properly extend the electricity supply to small enterprises internal and external to the camp. Examples of productive energy needs in camps include lighting, milling, refrigeration, powering small appliances, entertainment and communication/internet access.

Conclusion

The author argues that using market dynamics to deliver energy products and services is a viable option in most humanitarian situations, particularly in protracted crises as opposed to acute emergencies and also when preexisting delivery structures have not been completely disrupted by the crisis. When the opportunity is there, strengthening market systems is preferable to creating parallel delivery structures.

A market approach, as defined here, encompasses mixed models where services are delivered by private enterprises, but clients that cannot afford to pay receive subsidies increasing access and demand.

The level of involvement of the private sector can vary considerably according to the context and local conditions ranging from the involvement of local retailers engaged in food distribution through cash transfers to companies that provide energy to power operations in off-grid locations.

Letting systems develop organically through cash injections, service outsourcing and result based incentivizing mechanisms, could be a more flexible and effective way to bring change in the energy field and achieve the overall goals of sustained and sustainable development.

One barrier to the introduction and wider diffusion of these approaches is the humanitarian-development divide contraposing "saving lives" to laying the basis for longer-term development, as if the two were opposite and mutually exclusive aims. In reality, as argued in this chapter, the picture is much more blurred with markets surviving disasters or quickly returning in the immediate aftermath. Particularly useful, to enable quicker market-based response would be to include energy in the scope of work on preparedness seeking to introduce measures that will mitigate future disaster before it strikes.

Another challenge is the lack of specific knowledge and capacity, starting with more appropriate measuring and budgeting methods, to the ability to design market-based delivery models.

However, with more evidence coming out of pioneering experiences, renewed efforts to diffuse innovative solutions and the support of visionary humanitarian donors, there is a strong opportunity to start and sustain an energy revolution for the humanitarian sector.

Notes

1 The average lifetime of a refugee camp is estimated to be 18 years; for more details on this calculation see Grafham, O. and Lahn, G. (2018), 'The Costs of Fuelling Humanitarian Aid', *Chatham House*, London, UK.

2 Cash/Voucher definition: The provision of money or vouchers to individuals or households, either as emergency relief intended to meet their basic needs for food and non-food items or services, or to buy assets essential for the recovery of their livelihoods (DG ECHO, 2019).

3 M-Pesa is a mobile banking service that allows users to store and transfer money through their mobile phones. M-Pesa was originally introduced in Kenya as an alternative way for the unbanked population of the country to have access to financial services. Safaricom, the largest mobile phone operator in Kenya as of 2016, launched M-Pesa in 2007.

4 A list of available tools is given on the webpage of EMMA ("Emergency Market Mapping and Analysis Toolkit".

5 More information is available on the website of ENDEV.

6 REDD+ is a mechanism under the United Nations Framework Convention on Climate Change to reduce emissions from deforestation and forest degradation in developing countries.

7 While Alternate Current is commonly used in grid systems, Direct Current is used to charge batteries and as a power supply for electronic systems.

Bibliography

Arctic Energy Alliance (2011), 'Guide to Best Energy Practices for Remote Facilities', *Arctic Energy Alliance*, Canada.

Bastagli, F., Hagen-Zanker, J., Harman, L., Barca, V., Sturge, G., Schmidt, T. and Pellerano, L. (2016), 'Cash Transfers: What Does the Evidence Say? A Rigorous Review of Programme Impact and of the Role of Design and Implementation Features', *Overseas Development Institute*, London, UK.

Bellanca, R. et al. (2013a), 'Delivering Energy for Development', *Practical Action Publishing*, Rugby, UK.

Bellanca, R. et al. (2013b), 'Designing Energy Delivery Models that Work for People Living in Poverty', *CAFOD-IIED*, London.

Bellanca, R. (2014), 'Viewpoints: Interview with Raffaella Bellanca, Ewan Bloomfield and Kavita Rai', *Boiling Point*, 63.

Betts, A. et al. (2014), 'Refugee Economies Rethinking Popular Assumptions', Humanitarian Innovation Project, *University of Oxford*, Oxford, UK.

Cook, O., Forte, J., Mozersky, D. and Terrada, R. (2019), 'Peace Renewable Energy Credits Facilitating High-Impact Projects in Fragile Regions', *Center for Resource Solutions in Partnership with Energy Peace Partners*, San Francisco.

Corbyn, D. and Vianello, M. (2018), 'Prices, Products and Priorities Meeting Refugees' Energy Needs in Burkina Faso and Kenya', Practical Action research paper for the Moving Energy Initiative, *Chatham House*, London, UK.

DG ECHO (2019), 'Cash Transfers', Directorate General for European Civil Protection and Humanitarian Aid Operations, *European Commission*, Brussels, Belgium.

Endev (2018), 'Results-based Financing for Energy Access How to Design and Implement Projects: Lessons from the Field', *Deutsche Gesellschaft für Internationale Zusammenarbeit (GIZ) GmbH*, Eschborn, Germany.

Grafham, O. and Lahn, G. (2018), 'The Costs of Fuelling Humanitarian Aid', Chatham House research paper for the Moving Energy Initiative, *Chatham House*, London, UK.

Jagger, P. and Dasa, I. (2018), 'Implementation and Scale-up of a Biomass Pellet and Improved Cookstove Enterprise in Rwanda', *Energy for Sustainable Development*, 46 (October), 32–41.

Levine, S. (2017), 'Markets in Crises: The Implications for Humanitarian Action', Humanitarian Policy Group Report, *Overseas Development Institute (ODI)*, London.

Mastercard and Western Union (2017), 'Exploring Opportunities for a Digital Ecosystem in Kakuma and Kalobeyei', *Mastercard and Western Union*, Purchase, NY.

ODI (2015), 'Doing Cash Differently How Cash Transfers Can Transform Humanitarian Aid', Report of the High Level Panel on Humanitarian Cash Transfers, *Overseas Development Institute (ODI)*, London, UK.

Okello, S. (2016) 'The Energy Situation in the Dadaab Refugee Camps, Kenya', Energy for Impact research paper for the Moving Energy Initiative, *Chatham House*, London, UK.

Rivoal, M. and Haselip, J. (2017), 'The True Cost of Using Traditional Fuels in a Humanitarian Setting: Case Study of the Nyarugusu Refugee Camp, Kigoma Region, Tanzania', UNEP DTU Working Paper Series 2017: 3. *UNEP DTU Partnership*, Technical University of Denmark.

Rivoal, M. and Haselip, J. (2018), 'Delivering Market-Based Access to Clean Cooking Fuel for Displaced Populations the Kigoma Region, Tanzania: A Business Plan', *UNEP DTU Partnership*, Technical University of Denmark.

Vianello, M. (2016), 'The Energy Situation in Goudoubo Refugee Camp, Burkina Faso, Kenya', Practical Action research paper for the Moving Energy Initiative, *Chatham House*, London, UK.

WFP (2017), 'Pro-Smallholder Food Assistance: A Strategy for Boosting Smallholder Resilience and Market Access Worldwide', *World Food Programme*, Rome, Italy.

WFP (2019), 'Cash Based Transfers, January 2019 Update', *World Food Programme*, Rome, Italy.

7 The role of market systems in delivering energy access in humanitarian settings

The case of Burkina Faso

Mattia Vianello and Anoushka Boodhna

Introduction

This chapter explores the idea of delivering energy access through market systems and outlines why this is important, relevant and useful in humanitarian settings. The chapter reflects on the experiences of the Moving Energy Initiative (MEI)[1] in delivering a market systems development project for renewable energy in an area of Burkina Faso hosting a significant number of refugees. The project team observed the impact of this intervention and the possible lessons that have been generated for the humanitarian and development sectors. The chapter discusses the experience of the MEI project team in designing and delivering a project from assessment to delivery and the opportunities uncovered: from building trust between refugees and the host community, to the importance of market incentives and market facilitation tactics that bring about sustainability beyond any development initiative and encourage adaptation within the energy sector. The chapter closes with key thoughts on what should be considered when thinking about energy systems. The sub-Saharan Africa energy market remains very fragmented so an approach that strengthens market systems could be very useful for the humanitarian aid sector and beyond. However, this means changing the way humanitarian aid agencies deliver energy in displacement settings, including recognising that the conditions are often suitable for market systems development despite immediate perceptions.

Energy access and market systems

Energy is a fundamental need and this is no different for refugees. In particular, energy is needed to develop livelihoods, access health and educational services, and increase the quality of refugee lives. As earlier chapters have already made clear, energy access in humanitarian settings is generally funded by aid. But are there better ways of providing energy? Is there a need for better diagnostic tools? And do we also need to change our expectations of how energy can be delivered? What are the trade-offs between urgent response and longer-term sustainability?

Humanitarian assistance and aid delivery are changing. As Mach shows in Chapter 2, there is an increasing awareness within the sector that traditional aid approaches are insufficient. First, traditional methods of energy delivery by aid agencies can keep refugees on low-value energy solutions – tiers 0 and 1 of energy access related to Multi-tier Framework (World Bank, 2017). Second, they can create market distortions. Scoping studies carried out by MEI found that local enterprises operating in the areas surrounding refugee camps in Burkina Faso were having to compete with free products (Vianello, 2015). A proportion of energy products given to refugees as aid distributions often end up in the marketplace selling at very low prices creating unhealthy competition for local enterprises. This is because when aid agencies engage in a bulk distribution of products, like solar lamps or cookstoves, this will generally be undifferentiated and treat the refugee population homogenously. Refugees who do not need or use these products or prefer alternative products end up sending what they have been given to the marketplace – or they may repurpose them into something more useful (Cross et al., 2019).

Importantly, distributing solar lamps or cookstoves does not solve the refugee energy access problem (Bellanca, 2014; Lahn and Grafham, 2015). Although distributions might address some short-term need for a proportion of refugees, they patch over the deeper, systemic issues that have led to poor energy access in the first place – the supply and retail issues as well as the lack of supporting services, like credit, payment mechanisms and after-sales support, needed to make quality solar lamps accessible and useful for energy users. Moreover, aid agencies that engage in energy information delivery are insufficient. For example, one-off trainings on the harmful effects of wood fires overlook the deeper need for regular information flow on renewable energy, as well as the absence of any cooking alternatives that are cleaner, accessible and importantly, reliable. For example, in refugee communities, a shift from wood-burning to the usage of LPG requires the supply chains for LPG to work, and when it is unavailable, communities have no choice but to fall back to harmful wood-burning practices (Patel, 2019).

In order to address this, it has become pertinent to apply a development lens to refugee energy access – especially in the case of protracted crises. Doing so means thinking more about long-term planning instead of just short-term assistance, sustainability instead of dependency on aid delivery organisations and integration with the local context and national frameworks (and local enterprises) instead of stand alone, parallel structures.

As humanitarian actors increasingly adopt a development lens, the aid sector must also think more systemically and appreciate the complexity of how systems evolve. In humanitarian settings, the system is somehow broken and there are high levels of poverty, characterised by low incomes as well as low access to services that build capabilities. Such a systemic breakdown is caused by a number of interconnected factors, for example, remoteness from main economic zones and trading routes, conflict, violence and insecurity and climate-related effects, like drought or flooding. Addressing these multiple factors requires thinking about rehabilitating or strengthening broken systems for energy and strengthening the structures in the system to generate the launchpad for energy to take off, grow, adapt and evolve and meet the changing needs for local people. In areas with significant humanitarian presence, this take-off can be suppressed through the tendency to be short-termist and reactive to systemic problems and distorting markets and weakening market structures.

Large donors like DFID and USAID are increasingly focusing their investments in fragile and conflict zones on the need for systemic approaches inclusive of economic development and private sector engagement (see Ricigliano and Chigas, 2011; McIntosh and Buckley, 2015; Bhattacharya and Kharas, 2018). The role of the private sector in humanitarian assistance is also increasingly prominent. For example, World Bank and IFC recently published a report on the business opportunities in Kakuma refugee camp (IFC, 2018). Reports such as these indicate that market-based approaches could result in new job opportunities for refugees and the host community and improvement to service delivery, consumer choice, and contribute to self-reliance. So, rather than seeing refugees as passive recipients of aid, such studies are reconceptualising our thinking and helping us to understand that refugees can be more active agents of change, engaging equitably in market systems and being more integrated rather than isolated.

Private sector approaches in fragile zones can also be controversial because there is latent anxiety when thinking of generating profit from vulnerable people. When market systems are weak, this can open the door to unscrupulous behaviour, which can have an especially negative effect on vulnerable populations. Importantly, a key principle in systems thinking for the development of market-based approaches is how the formal and informal rules and norms in the market system promote and amplify 'good behaviour' and push out bad behaviour. These rules and norms govern the flow of goods, services and information and are generated by the patterns of behaviour within the actors in the systems. For example, in Sub-Saharan Africa, in energy markets for products like solar lamps, the performance of a product against what was promised through advertising and the role of consumer feedback mechanisms will be key to amplifying good performance and rooting out poor performers.

Thus, in a market systems approach, practitioners are primarily interested in working with actors who create value and have a growth mindset. Such an approach is agnostic to specific products and businesses and more interested in underlying behaviours and motivations. In this way, energy markets and private sector behaviour can be made more inclusive and equitable for vulnerable populations. However, this does not mean that humanitarian or development agencies should set rules and control behaviour to make systems safe for people – instead, agencies need to think seriously about creating the right underlying conditions for the system to be inherently inclusive. This might mean aligning incentives so that inclusion of marginalised populations is aligned with private sector performance.

Testing an energy market-system project in a humanitarian setting

There are important differences between 'market-based' and 'market-system' approaches. A market-based approach (broadly defined) includes any type of interventions based on market principles, particularly to facilitate access to energy products and services to meet the target groups' needs. A market system approach instead uses a system-thinking lens to diagnose the underlying constraints in the energy market where interventions are aimed at unlocking barriers to create functional markets. For example, in the scenario where a refugee is given grants or loans to acquire energy products through the market, this would fall within the category of market-based (refugees would purchase from firms). This is not necessarily 'systemic' because the aid agency is taking up a market function (providing loans, matching to firms). In this example, a more systemic intervention would be for the aid agency to facilitate the conditions for market actors – energy firms, financial institutions, etc. – to provide energy customers (refugees) financial support to purchase energy products possibly either as a product loan (financial institution gives loan to refugees to use for energy) or as embedded finance (energy firm provides finance to refugees in the form of credit, pay-as-you-go [PAYG], leasing, rent-to-own, etc.). This takes longer and requires thinking about small, targeted interventions to support firms and refugees to make incremental changes along a behaviour pathway. To this end, market systems specialists are interdisciplinary and use perspectives, tools and frameworks from a variety of fields to get to the underlying causes of dysfunction that matter to people living in poverty: from economics, to strategic business management, complexity science and social sciences.

Market exchanges are complex and simple solutions and technical fixes are insufficient

Poverty and systemic dysfunction are complex. For complex problems, simple technical fixes do not work. A classic technical fix in energy access for refugees might be to impose a specific course of action by distributing specific energy products (such as solar lamps) or to train refugees around the negative effects of wood and charcoal use. A systemic approach would be to work with complexity and not against it, which means working with norms, behaviour and preferences and the network of actors, exchanges and functions that exist on the ground. This also means thinking about the context – supply chains and market chains, laws, regulations, standards, social rules – that influence how this happens. This includes working with aid agencies themselves to challenge the idea that refugees can only tolerate handouts, that market systems do not exist and that market strengthening is pointless. This perception can result in the lack of motivation for more complex approaches and underline the preference for more simple fixes, further enforcing why market systems might not work well for the people within it.

The systemic approach to markets is also relevant for the strengthening of public services. Learnings in the health sector and water and sanitation can be applied to envisage how a market systems approach to energy could be relevant. In health, a three-year action research project funded by UKAid Kenya explored how markets could complement public health services, and where low-income target groups were paying for healthcare in different ways through private sector pharmacies, laboratories, health insurance providers, medical suppliers (like Glaxo SmithKline), health clinics and private community-based midwives (PSP4H, 2016). In addition, in water and sanitation, it is not simply about access to water wells and boreholes; it is about strengthening service provision in general, including small-scale private sector approaches, and building incentives for fee collection, life cycle costs of long-term operations and maintenance, spare part supplies, and storage of fees over a long period (Bourque and Mitchell, 2016). So, a market systems approach to strengthen energy as a public service would similarly involve working through a variety of market actors in a variety of ways.

There are emerging opportunities for energy market systems and refugees

In Burkina Faso, the MEI team worked in Goudoubo refugee camp in the north of the country in the Sahel region. The camp was established in 2012 and is one of three official sites, alongside Mentao and Djibo, that together house more than 24,000 refugees in Burkina Faso (in addition to over 81,000 internally displaced people) (UNHCR, 2019). Goudoubo itself has almost 3,000 households and around 10,000 people. Importantly, in Burkina Faso, in 2014, it was estimated that 43.7% of the population were living below the poverty line (World Bank, 2019). Burkina Faso is ranked 183rd in the Human Development Index (UNDP, 2018), it has a gross national income per capita of $620.14, with a population in 2018 estimated at more than 19 million people – 70% of whom live in rural areas.

The MEI market systems development project in Goudoubo camp, Burkina Faso, was one workstream of the MEI work, which built on three previous years of research. The aim of the project was to demonstrate the potential for the private sector to play a role in delivering energy to the camp. MEI initially collected data on energy needs and willingness to pay in Goudoubo camp providing a good understanding of consumer level and demand side issues. A sector overview report, Building Energy Access Markets, was used to learn about the energy access landscape and provide a framework for a deeper, contextualised exploration in Burkina Faso (Franz, 2014). Policy documents were reviewed to identify where future investment was headed and where policy could be influenced. Household surveys, key informant interviews and focus group discussions were carried out with refugees (as energy consumers and users) to identify the deeper needs and preferences of energy users, and key informant interviews and consultations were also carried out with the private sector (as energy suppliers and supporting functions) and humanitarian agencies (those involved in energy provision as intermediaries). A systems analysis of the energy market in Burkina Faso was then carried out.[2] For example, the MEI team used design-thinking research methodologies commonly found in modern, creative sectors, such as ICT, to explore ideas around what was needed for the energy system to work better and unpack the experience refugees had interacting with private sector and aid agencies. The findings highlighted what factors affect willingness and ability to engage in markets and pay for goods, services and information and what incentives exist for the private sector to build relationships with refugee communities now and in the future.

To better understand how the market system and its composite relationships, connections and structures could be strengthened, the MEI team looked at where the issues were cross-cutting and interdependent – for example, solar products and LPG supply were both affected by a lack of retail sector. The team also looked at where the system was already evolving and who was already demonstrating the incentives to engage in energy markets for refugee access – for example, the research found energy firms who had already engaged in sales and marketing in other rural areas of the country and were interested in expanding to provide energy for refugees in Goudoubo. The team then built the project based on five key insights related to interconnected markets, marketplace access, leverage points, adaptation and catalytic investments.

There is an overlap between the energy needs of refugees and the wider host community

Refugees are often settled in some of the most underdeveloped regions of their hosting country; especially in Africa, which hosts over 18 million refugees (UNHCR, 2019). This means that it is often difficult to separate the needs of refugees from needs of host community. An integrated approach to energy which considers refugees and hosts also reduces tensions between the groups and builds trust.

In Sub-Saharan Africa, in general, when the MEI team compared the percentages of rural host communities as compared to refugees who lack access to electricity or to cleaner fuels, it was evident that the two groups face similar energy access challenges. For example, the electrification rate in Burkina Faso is 19% and only 9% of the population have access to clean cooking solutions (World Bank, 2017). Those figures aren't very different from the situation in Goudoubo camp, where 99% of refugees have between tier 0 and tier 1 for both lighting and cooking (Corbyn and Vianello, 2018).

In Burkina Faso, Goudoubo refugee camp was located only 15km from the host community situated around the main trading town, Dori. Again there was significant overlap in the energy needs of refugees in Goudoubo and those of the nearby host community. Both groups: 1) relied heavily on firewood and charcoal as fuel for cooking, 2) had limited to no access to the national electrical grid, 3) were mostly unable to access alternative energy solutions because of an underdeveloped supply and retail network for both solar products and LPG canisters and cookstoves, and 4) were constrained by high levels of income poverty – although in fact, due to handouts from aid agencies, the average monthly income for refugees may sometimes be higher than that of the host community (Corbyn and Vianello, 2018). In Dori, the MEI team found that enterprises were already providing solar energy and local populations were increasingly using solar energy – meaning that both the supply side and demand side dynamics looked promising. So, Dori was selected as the key area of activity for MEI activities, primarily in order to act as a springboard to attract more market interest to the region and generate spill over effects to the camp.

Refugees are already accessing marketplaces in the wider host community

Certain groups of refugees from Goudoubo were able to travel – with specific permits – to Dori where they were able to engage in some sort of economic activity or get access to information. In Dori, there were functioning marketplaces delivering goods and services in sectors such as livestock, dairy and crafts. Livestock and dairy were the main socioeconomic activity; and craft activities included painting, fabric, basketry, mechanics and electricity. Trade flows were primarily constituted of local products including agro-pastoral products (cereals, oilseeds, livestock, milk, leather and skin, oil, canned tomato, rice), building materials and hydrocarbons (petrol, diesel and lubricants). Some refugees also traverse the 265km distance to the capital city, Ouagadougou. MEI also found the economic activities between camp residents and communities are intertwined, with refugees selling goods in the host communities and people from the host communities selling products in the camps (Corbyn and Vianello, 2018).

MEI was able to ascertain that refugees were interested in accessing energy services through the market. Refugees felt that through the market they would be able to get a better choice of energy products (the product offer was scarce at camp level, and often it was only undesired products that were available), and many refugees were interested and able to pay for better quality (many had been put off accessing solar energy because of product breakdowns in the past). MEI explored willingness and ability to pay and found that for a basket of goods (energy and non-energy), refugees were able to make some upfront payment and were interested in covering the remaining costs through payment facilities, like installments. Therefore, MEI knew that a project that appreciated this complexity of interconnected factors needed to be developed – and simply supporting energy firms to deliver products to the camp was not going to address the finance, information and quality needs of the system.

There is an opportunity to intervene at market-based leverage points

The perception is that the private sector is not interested or able to deliver energy goods, services and information to refugee communities; and it is true that humanitarian activities tend to occur in 'thin' markets – where there are very few firms and those that do exist experience difficulty finding and transacting with each other at viable costs. As a result, aid agencies tend to directly create supply chains by paying private sector operators to engage with the market through procurement contracts. However, in Burkina Faso,

MEI explored the wider system in which refugee camps were embedded and found that there were valuable leverage points. To find these, it was important to look for the existing, and possibly 'hidden', market activities and economic flows in the informal economy, which is particularly pertinent when working in highly underdeveloped zones. MEI worked with market actors – energy firms, financial institutions, training institutes – who were already operating formally and informally in Dori. In this way, MEI avoided heavily subsidising actors and instead focused on small, targeted, catalytic 'cost-shares' to help firms adapt parts of the strategy and enhance their outreach. MEI found there was no need to take up any transport costs for energy suppliers because there was already a transportation network in operation for food, household products, animals, etc., which energy firms could leverage, taking advantage of the vehicles used by refugees and others that make the daily commute by motorcycle or vehicle to and from Dori and Goudoubo camp.

Adaptation and expansion of energy systems meets the needs of the people within

In their conversations with the private sector, MEI found that using Dori as a springboard – and supporting energy firms to develop strategies for Dori as well as the camp – would offer enough market potential to be attractive. Primarily this was through sheer size of population, but it also helped firms to have access to important supporting services, such as the transportation networks and retailers needed for supply and market chains to work. MEI integrated the refugee camps and host communities. The economies of scale offered by looking at both refugees and surrounding communities as a whole can be multiple. Last mile distribution is expensive, and is often not viable when the consumers are spread around in small numbers across vast areas. Thus aggregation points – such as a refugee camp – represent an opportunity for market actors to leverage networks. In addition, aggregating energy usage and increasing load demand for energy infrastructure, for example, through mini-grids, means community services such as schools and clinics can be better served by solar hybrid mini-grids that could be used by both refugees and communities. Also, when load increases, more connections could be created to businesses or even households in the area. Private developers or government agencies can therefore explore models for electrifying those camps, taking into account the willingness to pay in refugee and host communities and designing suitable subsidies that could be covered by humanitarian agencies to support service delivery in the short term. Finally, a better value proposition for financial providers to support energy suppliers and consumers is created and adapted business models for some energy services can emerge (like LPG supply and refill). These make financial sense only when a critical mass of consumers exist.

MEI knew that over time, energy supporting functions would be more likely to take root and evolve in a place like Dori with outreach structures to the refugee camp. The key was to 'seed the right conditions' for energy systems to sprout and take root. MEI wanted to support a vision for how an energy system could meet the needs of the host population as well as energy users from refugee camps. As the population increases and changes, the system would be able to adapt and expand. It is better for there to be the conditions for new parts of the system to emerge than to think about how to deliver more aid and new energy access projects each and every time the town or refugee camp expands. This in-built sustainability and scalability aspect of market-systems development is crucial.

Targeted, catalytic investments facilitate the emergence of better functioning energy markets

In traditional projects, there is often a tendency to deliver large-scale complicated projects to target every component of the complex problem. Some projects deploy a series of best practice manuals to 'fix' problems; others set up a set of complicated contracts. MEI instead focused on an approach that was designed to be flexible and 'shift' with the dynamics of the system, allowing for local adaptation.

Following the analysis, MEI supported marketing 'learning launches' for energy firms to help them identify the market they were targeting and what market development strategy they wanted to start with. MEI supported market interventions that gave refugees and private sector practical experience of interacting with each other. To be adaptive, MEI did not make demands for long complicated work plans and instead iterated activities based on market feedback. MEI also did not focus simply on increasing the sales of energy products or revenues generated for the firms. Although sales metrics might be relevant, MEI focused on the quality of the relationships between actors in the system. Over time, MEI observed changes in the interactions between market actors – especially for poor energy users in host communities and refugee camps. In some cases, it can be said that these interactions moved from one-off, sales-driven transactions to more value-driven, relationships.

In practice, this meant supporting the following activities: 1) field visits to Dori and Goudoubo for energy firms, capturing information and market insight through a variety of media such as video; 2) networking events to bring energy firms together, as well as supporting services and direct marketing activities (including a trade fair, a mobile caravan and motorbike sales); 3) capacity building of marketing agents – directly funded by the firms – so that firms had a more long-term presence in the market and refugees had an entry point to follow-up on products; 4) relationship-building between firms and public services, such as the local mayors' office, the radio station and a solar technician training institute; 5) relationship-building between finance and energy firms including the exploration of pay-as-you-go and other fintech opportunities; and 6) adaptations to local curriculum to improve the availability of trained solar energy technicians in Dori and Goudoubo including the roll out of cost-shared repair clinics in Goudoubo camp.

Over 12 months, MEI was able to demonstrate that a market systems development approach for energy and refugees could make significant improvements to the way that refugees accessed energy. MEI was able to support two energy firms to set up long-term operations in Goudoubo and Dori. The companies themselves tested three direct marketing activities and embedded training activities for 48 marketing agents and seven energy technicians. The units sold through marketing activities came to approximately 215 pieces reaching energy users in Goudoubo and Dori. MEI did not need to subsidise the price of energy products. Importantly, beyond the marketing, MEI facilitated connections between up to 50 energy actors from private sector, NGOs and government. In particular, the MEI team catalysed discussions around the development of the LPG sector and fintech solutions for energy financing and solar home systems. Even once the project had ended, the actors continued to work together to identify opportunities to develop the market – sales of energy products have since doubled; the training institute is offering specific training for people living in Goudoubo camp and in the host villages, at least two technicians have set up a business delivering repairs for solar products; and one energy firm has been developing their own partnership with a local credit union to bundle finance with their energy products.

Lessons for the future

Taking a systemic approach beyond the camps

In the last decade the private sector has played an increasing role in delivering energy access in developing countries, for example, about 360 million people have accessed better energy services through solar technologies distributed by the private sector since 2010 (Dalberg Advisors and Lighting Global, 2018) – some of the key success factors have been the development of the 'last mile' distribution function, after-sale services, enabling policies and norms and financial mechanisms, such as pay-as-you-go and mobile money.

In a functioning market system, the vision is for energy users in humanitarian settings to have their energy needs met through the local system rather than continue to rely on handouts from aid agencies, donors or the Government (which might be unsustainable in the long run). Even in 'thin' or highly underdeveloped markets, there is an opportunity to potentially alleviate the constraints and help the market system grow and develop. However, this approach does not exclude the national authorities playing a role, which is fundamental in terms of creating the right conditions (through regulations, infrastructure, etc.), and also in providing the right financial mechanism for those who could not afford energy services even if they were accessible.

Of course a number of challenges remain. Market systems development might be more useful for protracted displacement situations than for an emergency response to crisis when immediate needs must be met rapidly. Some refugee camps are there for a generation, and the average age of a refugee camp is 18 years (Grafham and Lahn, 2019), with no visible end to conflict, so this is not a small subsection of the current cases. Many actors also struggle to navigate the complex bureaucratic systems of humanitarian settings and their complex regulations (Grafham and Vianello, 2018). In many countries, refugees' movement is restricted, which limits their ability to access markets where they can acquire energy products.

Despite this, there seems a good reason to question the typical paradigm where humanitarian organisations only focus on their target group and where energy interventions and government-led plans neglect those who fall outside national strategies, such as displaced people. When looking at the history of energy access programmes in the developing world, it seems that achieving scale in energy access programmes in Sub-Saharan Africa happens when aid agencies have played a more facilitative role and initiatives have been led by a mix of government-led programmers (for example, Practical Action illustrates this in the Poor People's Energy Outlook 2018 with the cases of clean cooking in Indonesia and rural electrification in Nepal) and the private sector, for example, in the SHS market, which has seen huge successes in the last decade (Dalberg Advisors and Lighting Global, 2018). Even in the latter case the existence of an enabling environment and the right conditions for market players to emerge came from the government. Energy access programmes, particularly those at scale and driven by governments, have often focused on supply, technology and infrastructure, frequently overlooking social and economic differences which exist also at the last mile (Practical Action, 2018). Newer approaches which include building markets for off-grid communities have shown significant promise and may perform better on inclusion. As displaced people are marginalised communities which have been historically left out of energy planning, there seems to be an opportunity for supporting the conditions for market actors to access those areas. A systemic analysis can therefore create the conditions where access to all forms of energy access for refugees can be achieved through a combination or private, public and humanitarian approaches.

Creating conditions for a market system approach

By looking at MEI and others' experience, it is possible to indicatively define the types of scenarios when a market system approach is relevant and for which there are sufficient conditions. Whitehouse synthesises those indicators of when a systemic approach can be the most suited mechanism: 1) protracted displacement – refugee camp there for a number of years; 2) existing energy access market opportunities in that country; 3) no need for urgent response; 4) existing economic activities; 5) funding allow longer-term results; and 6) conditions for shaping behaviour of other humanitarian organisations in the area (Whitehouse, 2019).

Table 7.1 from Whitehouse provides useful questions to consider when assessing those indicators and the suitability of this approach.

Table 7.1 Contextual influences for market selection and intervention design

	Contextual influence	*Market sector selection and intervention design requirements*
Crisis Type	Market system exists but in a state of failure or protracted crisis	Expected timeframe of displacement; affect of crisis on market functioning
Energy market maturity	Adaptability and risk appetite in the market	Market actor presence; market demand and supply dynamics; historical/political/economic issues affecting the market
Pop. size	Feasibility of the market for public or private sector investment	Energy demand host vs refugee community; inequities in energy access
Location	Supporting functions and rules required for market system	Market constraints; interactions between refugee camp and host community, and rural and urban spaces
Political/cultural reception	Rules of the system; role of humanitarian intervention; potential for market distortion	Incomes and willingness and able to pay; refugee reliance on aid and cash transfers
Aid influence	Perceptions regarding aid vs market viability	Dominance of the humanitarian presence; expectations of market actors

Source: Whitehouse, K. (2019)

Developing market-systems approaches for inclusion of the very poor or vulnerable, marginalised

As in any approach involving the private sector, a question remains on how to balance the incentives of firms, with the affordability of their products. Energy is a basic service, and a minimum standard of energy access is a basic service. In the case of rural energy the real cost of services can often be too high for the poorest of the community. In order to achieve full inclusiveness governments therefore need to work together with private actors to create the right conditions for both firms and consumers to meet at a satisfactory price point. This issue can be particularly relevant in rural electrification programmes in Sub-Saharan Africa, and even more when looking at refugee communities. In the displacement settings discussed in this book, a key stakeholder is the humanitarian sector. It is important therefore to explore how the humanitarian aid sector can work together with the private sector and the government to ensure that everyone has access to a minimum level of energy, in a way which is economically sustainable and is inclusive. Globally, there are beginning to be signs that institutions are recognising the limits of solely private sector roll out of off-grid electricity access as key organizations such as the SEforAll and World Bank, SEforAll and others, are recognising that some forms of smart subsidies and strong investment of public money will be needed to encourage the private sector to reach markets it currently sees as unprofitable (PPEO, 2017). It is therefore a good moment to bring this question to the humanitarian sector as well. Opportunities may exist for humanitarian organisations to play a role between the government and the private sector to ensure full inclusiveness to all.

Conclusion

This chapter outlines the details of a market systems approach applied to energy access in humanitarian contexts. The underlining dynamics of energy market systems were presented and the need for in-depth systemic analysis before designing any solutions was demonstrated. The chapter also illustrated examples of interventions with the private sector that can trigger a market without creating dependency or distortion. The linkage between energy markets for refugees and surrounding communities emerged as a key opportunity – both for private sector engagement and market systems development, as well as for rural energy planning. There now arises an important shift in practice and an opportunity to think more innovatively about how rural electrification plans led by governments can be integrated with private sector and market systems approaches and include displaced people living in camps. This ultimately raises the main question: what does this mean for the future role of humanitarian aid agencies organisations, and does thinking differently about energy systems for displaced people also catalyse a new wave of thinking about access to renewable energy in general, for underserved regions and poor populations all over the world?

Notes

1 The Moving Energy Initiative (MEI) is an international consortium seeking to change how the humanitarian system responds to the issue of energy. This chapter presents a case study also summarised in a Learning Brief published on the MEI website called 'Pioneering Market Systems for Energy Access in Humanitarian Settings – The Case of Burkina Faso'. Although, at the time of writing, it was still too early to assess the full impact of that work, there are already lessons for the sector. We focus on interventions in camps in Sub-Saharan Africa, but we find that lessons around system analysis and market system can be useful to a wider range of humanitarian contexts. The authors are the programme team who implemented the market systems development activities for MEI namely, Mattia Vianello, an energy access expert and the regional director for West Africa at Practical Action and Anoushka Boodhna, a market-systems specialist.

2 Some of the methodology used for the market-systems analysis is available on Beam Exchange, an online network and resource for market-systems development practitioners.

Bibliography

Beam Exchange (2014), 'Market Systems Approaches: A Literature Review', *Beam Exchange*.

Beam Exchange (2016a), 'Impact Evaluations for Market Systems Programmes', *Beam Exchange*.

Beam Exchange (2016b), 'Sustainability, Effectiveness and Scale', *Beam Exchange*.

Bhattacharya, A. and Kharas, H. (2018), 'Systems Thinking for Multilaterals: New Report on Global Financial Governance', *Brookings*, New York.

Boodhna, A. and Vianello, M. (2018), 'Pioneering Market Systems for Energy Access in Humanitarian Settings – the case of Burkina Faso', *Chatham House*, London, UK.

Bourque, R. and Mitchell, F. (2016), 'Turning the Tides: A Systemic Approach to Intervention in the Water Sector', *Enterprise Development and Microfinance*, 27 (1).

Corbyn, D. and Vianello, M. (2018), 'Prices, Products and Priorities: Meeting Refugees' Energy Needs in Burkina Faso and Kenya', Practical Action Report of the Moving Energy Initiative, *Chatham House*, London, UK.

Cross, J. et al. (2019), 'Energy and Displacement in Eight Objects: Stories in Sub-Saharan Africa', *Chatham House*, UK.

Dalberg Advisors and Lighting Global (2018), 'Off-Grid Solar Market Trends Report 2018', *International Finance Corporation*, Washington, DC.

Franz, M. (2014), 'Building Energy Access Markets: A Value Chain Analysis of Key Energy Market Systems', *EUEI PDF*, Berlin, Germany.

Grafham, O. and Vianello, M. (2018), 'Energy for Displacement: Key Issues', Energy for Displacement "Understanding the Challenges" ', Heed Briefing Paper No.1, *Scene Connect.*

Green, D. (2016), 'How Change Happens', *Oxford University Press*, Oxford, UK.

IFC (2018), 'Kakuma as a Marketplace', *World Bank Group and International Finance Corporation,* Washington, DC.

Lahn, G. and Grafham, O. (2015), 'Heat, Light and Power for Refugees: Savings Lives, Reducing Costs', Chatham House Report for the Moving Energy Initiative, *Chatham House*, London.

Meadows, D. (2017), '*Academy for Systems Change*', US.

MEI/E4I Katie Whitehouse Paper.

McIntosh, K. and Buckley, J. (2015), 'Economic Development in Fragile and Conflict-Affected States: Topic Guide', *GSDRC, University of Birmingham*, Birmingham, UK.

PSP4H (2016), 'Programme profile PSP4H: Private Sector Innovation Programme for Health'.

Ramalingam, B. (2015), 'Aid on the Edge of Chaos: Rethinking International Cooperation in a Complex World', *Oxford University Press*, Oxford, UK.

Ricigliano, R. and Chigas, D. (2011), 'Systems thinking in Conflict Assessment: Concepts and Application', *USAID's Office of Conflict Management and Mitigation (DCHA/CMM) and AMEX International.*

Sullivan, T. (2011), 'Embracing Complexity', *Harvard Business Review*, Brighton, MA.

Thoday, K., Benjamin, P., Gan, M. and Puzzolo, E. (2018), 'The Mega Conversion Program from kerosene to LPG in Indonesia: Lessons Learned and Recommendations for Future Clean Cooking Energy', *Energy for Sustainable Development*, 46, 71–81.

UNDP (2018), 'Human Development Report', *United Nations Development Programme.*

UNHCR (2019a), 'Fact Sheet: Burkina Faso (January 2019)', *The Office of the United Nations High Commissioner for Refugees*, Geneva, Switzerland.

UNHCR (2019b), 'UNHCR, Africa', *The Office of the United Nations High Commissioner for Refugees*, Geneva, Switzerland.

Vianello, M. (2016), 'The Energy Situation in Goudoubo Refugee Camp, Burkina Faso, Practical Action Report of the Moving Energy Initiative', *Chatham House*, London.

Whitehouse, K. (2019), 'Adopting a Market-based Approach to Boost Energy Access in Displaced Contexts', paper for the Moving Energy Initiative, *Chatham House*, London.

World Bank (2017), 'The Global Tracking Framework Report 2017', *World Bank*, Washington, DC.

World Bank (2019), 'Poverty and Equity Database', *World Bank*, Washington, DC.

8 Vulnerability of women and girls in refugee settings

Considerations for energy programming

Tamsin Bradley and Katherine Liakos

Introduction

This chapter shows how and where violence against women and girls (VAWG) intersects with energy programming in the specific context of Kakuma refugee camp in Kenya. The chapter argues for the value in applying a VAWG-sensitised lens to development approaches, including those not focused solely on ending VAWG, in order to make important inroads into making the lives of women and girls safer and more secure from violence. The links between VAWG and energy usage tend to focus on highlighting risk factors rather than generating a detailed, nuanced understanding of how different forms of violence intersect with poverty and fragility in refugee settings. By risk factors we mean activities that can increase vulnerability to violence, such as collecting firewood or walking at night in poorly lit areas, which are discussed in detail later in this chapter. We know that women and girls are at increased risk of violence in conflict and humanitarian crises due to displacement, the breakdown of social structures, a lack of law enforcement, the potential further entrenchment of harmful gender norms, and the loss of livelihood opportunities for both men and women in the community, among other reasons (see Murphy et al., 2016). The focus for this chapter then is to explore – through a review of the literature and new qualitative data – how energy and VAWG intersect both positively and negatively. The purpose is to better understand the potential for improved energy access and usage to help contribute to ending forms of violence.

It is important to note that the types of violence women and girls are most vulnerable to at times of conflict appear to map to those in non-emergency settings, namely IPV (intimate partner violence). A review by Murphy et al. (2016) found an estimated prevalence of sexual violence among refugees and displaced persons in complex humanitarian emergencies to be 21.4%. Another research brief which reviewed screening of gender-based violence (GBV) in humanitarian settings estimated similar numbers: "at least 1 in 5 refugees or displaced women in humanitarian settings have experienced sexual violence. This figure is even higher when considering women who experience violence or threats of violence by intimate partners or other family members" (IRC, 2015).

In reviewing what research exists linking VAWG, energy usage and programme design, it is clear that the complexity and high prevalence of different forms of VAWG in humanitarian refugee contexts are often missed. There seems to be a lack of awareness into the ways in which existing energy use heightens the insecurity of women and girls. By adopting a more VAWG-sensitive lens it is possible to identify simple measures that can potentially have a significant impact in reducing feelings of insecurity and actual instances of violence. This chapter does two things; first, it will present a review of what we know about the links between VAWG and energy, noting that most of the literature and programme focus has been placed on reducing firewood use as a means of mitigating the perceived risks associated with women and girls collecting firewood in camp settings. Second, it will present new qualitative evidence from 20 in-depth interviews with a number of female refugees in the Kakuma camp and Kalobeyei settlement in Kenya carried out in November 2017. This data reveals the horrifying extent of violence against women coupled with intense feelings of constant insecurity. These experiences and feelings are made more severe by the lack of lighting both in the home and in public spaces, the reliance on firewood as the primary fuel source and generally poor living conditions. Many other factors render women vulnerable to different forms of violence but findings (outlined in more detail throughout this chapter) suggest that improved energy, lighting and cooking fuel would significantly enhance feelings of security and may even support greater resilience to violence. If energy access programmes have the potential to do both good and harm, then we need to approach energy access work with an increased sensitivity to VAWG and proactively seek to contribute towards the goal of reducing VAWG. This chapter will evidence the need for greater consideration of VAWG in energy programming in refugee settings. It will present a VAWG mainstreaming lens as a useful mechanism for supporting greater awareness in the humanitarian energy community. The chapter is structured as follows: in the first section we review what is known in the international literature about violence against women and girls; the second section explores the energy and VAWG nexus in both the literature and our study; the third section focuses on outlining a recommended approach to mainstreaming a VAWG perspective in energy programming.

What do we know about violence against Women and Girls?

There is agreement that violence against women and girls is driven by inequitable gender norms, attitudes and behaviours relating to gender roles, family structures and responsibilities and the inequitable development and application of legislation, leading to the normalisation and consequently the perpetuation of violence (Duvvury and Scriver, 2014, p. 13). Violence should be seen as a cycle combining gender norms and attitudes, undervaluing of women and girls, disempowerment in public and private life, vulnerability to violence and finally normalisation of violence. In order to adequately conceptualise violence against women as "a multifaceted phenomenon grounded in an interplay among personal, situational and socio-cultural factors", Heise built on ecological models of violence (Heise, 1998, pp. 263–264). The model distinguishes between various personal and situational/

social levels in order to identify the ways in which vulnerability to violence may be intensified at each level of the model. This model is now widely applied by the research community as a means of mapping the complexities of violence. However, there is arguably a substantial disconnect between the theoretical understanding of VAWG and programming to end it, thus illustrating the difficulty of translating theory into practice.

The challenge to VAWG programming is that in order to effectively position actions to end violence against women deep contextual insights are needed into how violence is maintained and normalised at many levels from individual attitudes and behaviour though to poor justice infrastructure and low government commitment. Identifying entry points for change is problematic. However, as we will go on to explore, improved energy options may well present one possible focus that could at least make improvements in how refugee women and girls perceive their safety. Currently VAWG-specific programmes are underpinned by the need for behavioural change on the part of potential perpetrators (Oxford Policy Management, 2014), with an understanding that unless the behaviour of perpetrators is changed VAWG will not end. The importance of including efforts to change mindsets is not always well understood in the energy sector, making the bringing together of VAWG and energy programming fruitful in terms of pushing the boundaries of established practice in the energy field. Furthermore, available evidence suggests that the most successful VAWG-reduction programmes are multifaceted and actively engage key community members such as faith leaders and elders (Murphy et al., 2016). Risk reduction activities, so it is thought, will not make any inroads into this process. Perpetrator behaviour is legitimated by certain gender-societal norms that render women and girls inferior and vulnerable to violence. In the VAWG community, violence is essentially about power imbalance (Kelly, 1988; Heise, 2011). The evidence to support this claim is strong and future energy programmes which emphasise their VAWG credentials need to take this into account more fully. Nonetheless, as we go on to argue, living in constant fear of violence erodes at a women's self-confidence, well-being and resilience (Le Masson et al., 2018). Any measure that could mitigate these feelings of insecurity would improve day to day living conditions and perhaps support collective action to challenge violence.

The VAWG and energy nexus in development and humanitarian contexts

As discussed previously, the reasons for an increase in VAWG during conflict and humanitarian crises are complex and multifaceted. The relationship between VAWG and energy access in humanitarian contexts tends to be defined by the increased exposure to VAWG risk factors, rather than any of the intrinsic causes. Fulu et al. (2013) argue that violence is driven by gendered social norms and a complex interplay of individual, community, and society-level factors, many of which apply in humanitarian and displacement contexts. The increase in exposure to risk is both a result of the prevailing energy access situation, in particular relating to firewood collection and camp lighting, as well as reduced privacy and security in camps and displacement settings (Listo, 2018). This increased exposure to risk may provide an opportune setting or location for GBV, but importantly it should not be regarded as its cause (CASA Consulting, 2001).

Potential risk factors and feelings of insecurity are frequently noted by women and girls in camp settings, most commonly in relation to poor lighting in public spaces and the need to collect fire wood (Corbyn and Vianello, 2018). These issues also strongly emerged in the qualitative research presented later in this chapter.

Cooking and firewood

There is very little data available on the overall humanitarian energy situation but it is clear that few refugees or displaced people have access to modern energy services. In most cases households rely on firewood for cooking. Estimates by the Moving Energy Initiative have concluded that "80 per cent of the 8.7 million refugees and displaced people in camps have absolutely minimal access to energy, with high dependence on traditional biomass for cooking and no access to electricity" (Lahn and Grafham, 2015). Although agencies often supply some firewood (or other cooking fuels at some sites), the bulk of wood must be collected or bought. Firewood collection is primarily carried out by women and girls and involves trips outside the camp. The frequency, distance travelled and length of each trip depends on the demand for fuel and local fuel availability.

The incidence of VAWG when venturing outside the camp settings is well documented in a range of literature on gender and energy access. A recent rapid review carried out for DFID's VAWG helpdesk identified some of the key evidence available on the links between access to different sources of energy at the household and community level and VAWG in camp-based settings (Parke and Fraser, 2015)[1] (see also Gunning, 2014). Examples of incidence of VAWG outside camps include Lahn and Grafham (2015); WRC (2006); and Mercy Corps (2010). VAWG is difficult to measure and is greatly under reported for a number of reasons including the lack of safe reporting mechanisms, lack of support and fear of being ostracised or punished by their communities. In addition, where firewood collection is illegal outside camps, this can further encourage exploitation and under reporting.

Although the increase in exposure to GBV risk factors is associated with leaving the camp, not with the firewood collection per se, firewood collection can mean venturing to more remote and isolated locations. Trips outside camp can also be for water collection, selling/buying goods, work, grazing and grass collection. However, many of the cited occurrences do reference firewood collection rather than other reasons.

In addition to protection risks in travelling to get firewood, assessments by the World Food Programme (WFP) have highlighted other risks in accessing fuel including domestic violence when there is no fuel for cooking and transactional sex in exchange for cooking fuel or for money to purchase it (WFP, 2013).

There are some anecdotal claims that GBV incidence has been reduced following changes to stoves, but no evidence to support statements. It can also be difficult to attribute changes to one project since changes in behaviour or reported incidences of violence could also be a reflection of factors outside the energy project. For example, a WFP SAFE cookstove project at Kakuma, Kenya, revealed some anecdotal evidence that there was a reduction in domestic violence due to less undercooking and skipping of meals and that some transactional sex may have been mitigated due to the new efficient stoves (WFP, 2013). Similarly, an evaluation of the Doro refugee camp in South Sudan (Danish Refugee Council, 2012) reports on how firewood and fuel are linked with some forms of VAWG, including domestic violence when food is scarce or undercooked, and sexual exploitation due to poverty. Rather than being a cause of violence, insufficient firewood and fuel is linked with experiences of poverty, which heightens the risk of men using gendered violence (Listo, 2018).

New stoves and reduced firewood use do not always result in a reduction in fuel collection trips. For example, in Darfur women still collected wood for sale after the distribution of improved cookstoves. Some studies even suggest that the time savings have fuelled the growth of a secondary market for the sale of firewood (Abdelnour, 2010). It is also important to note, as earlier, that travel outside camps is not restricted to firewood collection.

The review titled 'Gender-Based Violence in Humanitarian Settings: Cookstoves and Fuels' produced by the Global Alliance for Clean Cookstoves (August 2016) studied 15 humanitarian energy projects that included reducing GBV as an objective. It provided the following findings. Reducing the need to collect firewood does not necessarily bring down the total number of rapes. The most prevalent form of VAWG (as already stated) is in fact IPV not rape in public by a nonpartner (which most firewood rapes are classed as) or also conflict rape. Evidence capturing even a decrease in rape as a direct result of firewood collection is flimsy at best, particularly due to a lack of baseline and end-line data collection for humanitarian energy projects. Additionally, there is a lack of understanding around which cooking interventions are likely to have the strongest protection impacts. For example, only providing firewood that meets household cooking needs may not address all GBV related to firewood collection.

Whilst a reduction in risk does not automatically lead to a reduction in VAWG, it should be acknowledged that the collection of firewood presents a significant danger to women and girls. It is also clear that at the very least women and girls are fearful of being raped while undergoing this activity. The knock-on impact this will have on the psychological well-being of women and girls – regardless of actual harm through rape – should not be ignored.

Lighting

Protection issues, due partly to lack of lighting (along with privacy issues), are frequently cited as barriers to the use of the communal WASH facilities and as the rationale for fears of using the communal kitchens. For example, at the Zaatari camp in Jordan, women identified the WASH facilities and kitchens as the most insecure locations and 24% of participants said that safety concerns stopped them using communal kitchens (Serrato, 2014). Women feel unsafe and incidents of VAWG are reported to take place in poorly lit areas. IOM carried out two rapid assessment surveys and found that GBV prevalence was very high in two internally displaced people (IDP) settlements in Somalia and that many of the incidents took place at night in the dark. Safety issues are affected by lack of lighting and also by design and the distance to the facilities. Poorly designed lighting interventions could increase protection risks. For example, illuminated communal areas may encourage people to leave their homes at night to walk through poorly lit areas (Perkins, 2015). Research by the Moving Energy Initiative at Goudoubou refugee camp revealed that 'security and safety' was the first reason given by respondents regarding the importance of electric lighting (Corbyn and Vianello, 2018). At the time of this research there were no street lights in Goudoubou though it was cited by refugees as a priority for camp improvements – 86% of survey respondents (male and female) said that more household members would go out after dark if there were better public lighting (Corbyn and Vianello, 2018).

Although there is a link between lighting (handheld lanterns and street lighting) and reducing VAWG risk factors, there is little evidence since there are no studies that have monitored the incidence of VAWG before and after the introduction of a lighting initiative. One study tried to assess the effectiveness of solar lamps to reduce insecurity and SGBV (sexual and gender-based violence) in Puntland, Somalia, but the low reporting levels of such cases of violence rendered the task almost impossible (DFID, 2014). However, that study and other surveys and assessments do show that women often feel safer and appreciate the lights, as well as reduced fumes and risk of fire related accidents that can occur with candles and kerosene lamps (Corbyn and Vianello, 2018). For example, following the pilot launch of its Light Years Ahead initiative, UNCHR carried out a survey in three countries, with the findings showing that 60% of respondents feel safer using the bathroom at night (UNHCR, 2012). Research commissioned by the Moving Energy Initiative in Kakuma I camp highlighted that torches were the preferred method of lighting, though as a single-point task light they offer limited illumination for a family (Corbyn and Vianello, 2018). It should be noted that women and children typically have second-rate access to household lights according to Rom, Günther and Harrison (2017, cited by Corbyn and Vianello, 2018) which emphasises the importance of consider household power dynamics and the precise needs of both male and female end users.

However, a study undertaken by International Rescue Committee (IRC) which evaluated the impact of the distribution of handheld solar lights on women's and girls' perceptions of their own safety in two Haitian camps found that women did not feel safer (IRC, 2014). Before and after comparisons found that although 95% of the women and girls reportedly used the lamps, the women's own perception of their safety remained the same or worsened (largely attributable to other broader security issues). The lamps did not address the women's primary safety concerns (generalised crime, violence and mistrust, including sexual violence and harassment). In another case in the Democratic Republic of the Congo, UNHCR have received feedback of an unforeseen consequence that IDPs do not use their solar lamps at night for fear of revealing their location to rebel groups (IOM, 2011).

The IRC study concluded that handheld solar lamps are an important personal resource for women and girls however the root causes of VAWG are complex and cannot be addressed by a stand alone distribution of lamps.

VAWG – energy nexus in Kakuma camp and Kalobeyei settlement in Kenya

We will now move on to explore the issues and linkages raised earlier through new qualitative data. As stated in the introduction, we conducted twenty in-depth qualitative interviews with female adult refugee inhabitants of the Kakuma camp and Kalobeyei refugee settlement in Kenya, as well as two focus groups. We were helped in our sampling by staff working for various UN agencies. They helped us identify women from different ethnicities and cultural contexts and in different age groups. We acknowledged from the outset that these settings are different, and the protection procedures and issues also differ. We wanted to capture these differences in order to pinpoint any triggers or pressure points linked to location, history and ethnic mix of each context. The purpose of the interviews was to interrogate how and if energy usage feeds into VAWG either positively or negatively. Additionally, ten key informant interviews with stakeholders were undertaken in order to map out both energy programming and interventions that address violence against women and girls. This qualitative research provided in-depth insights into delicate topics that large-scale survey approaches cannot capture. The value of this approach is that it makes visible the complexities involved in delivering services and can therefore capture any negative and unintended consequences (or likelihood of these). Qualitative approaches generate a better understanding of the various relationships, values and beliefs that may enable or prevent access, in this case, to better more sustainable energy. Understanding the prevailing/dominant cultural and social norms is critical for issues of access and usage to be understood. Twenty interviews were sufficient to present a detailed picture and reach saturation of data collected, meaning that similar patterns in responses were documented across participants with little variation.

Background to Kakuma camp and Kalobeyei settlement

The Kakuma camp was opened in 1991 by UNHRC in response to the migration of unaccompanied minors from Sudan who had fled from conflict there. The camp has grown from 8000 when it first opened and is now around 149,374 (UNHCR February 2019 figures). Following a continuous influx of South Sudanese refugees after renewed conflict broke out in South Sudan in December 2013 the UNHCR requested additional land for expansion of the Kakuma Camp. In June 2015 the Turkana County Government allocated a site near the Kalobeyei Township, situated 40 kilometers Northwest of Kakuma and measuring 15 square kilometers.

The main demographic make-up of Kakuma, according to data from UNHCR February 2019, is:

- South Sudan 54%
- Somalia 23%

- Democratic Republic of Congo 7%
- Sudan 6%
- Burundi 5%
- Ethiopia 4%
- Uganda 1%

The findings are given later under the key themes that emerged. In summary, 100% of the participants had experienced some form of violence at some point in their lives. In many ways our findings map to what is known in the global literature. First that IPV is the most common form of violence even in humanitarian contexts. Our findings also highlight how different forms of violence link together. Whilst our data revealed that there are risks and insecurities associated with poor energy, light and cooking methods in particular, removing these will not trigger an end to all forms of violence. That said, it is clear that focus on improving specific energy needs will reduce insecurities and promote greater well-being in women and girls. This in turn may well translate into greater willingness and mental capacity to challenge violence.

Intimate partner violence

According to our findings domestic violence, in all its forms, physical, emotional, psychological and economic was reported to be the most common form of violence experienced in both Kakuma camp and Kalobeyei settlement. The following passages offer examples of the type of pressures and harrowing experiences many women endure.

> *I was a victim from my ex-husband who was beating me every time and he was drug addicted and he would chew miraa and every night after chewing miraa he would come and beat me and he turned the children against me and told the children not to help their mother and then later on he assaulted me badly whereby I still have injuries on my head and around my neck and other parts of my body. Later on he divorced me and after he divorced me he stopped supporting the children and he is now harassing me and he is going around the camp saying I have other men and he is also inciting the children and asking them if there are other men visiting the home so that is now the harassment which is going on. It has made me have psychological and emotional problems and sometimes when I go back I cry alone and it's disturbing me.*
>
> *Participant, Kakuma Camp*

The participants stated that IPV increases and is triggered by the pressures of living in the camp or settlement, poverty and no income. Some of the women interviewed reported that husbands regularly accuse wives of affairs and dressing inappropriately.

Research exploring links between hegemonic constructions of masculinities and violent male behaviour towards partners finds that disruption to the image of a man as a breadwinner spikes increased aggression. Other triggers identified in global literature include alcohol and drug abuse, and accusations that wives are not sufficiently fulfilling their domestic duties. In a refugee setting these triggers amplify. Very few of the women we spoke to had husbands earning a wage and stated that this had turned into aggressive behaviour towards them. A lack of fuel wood for cooking was also mentioned as a source of tension and aggression with husbands accusing their wives of not looking after them.

IPV is also perpetrated by the vulnerabilities felt by single women. In the Kalobeyei settlement there are many female-headed households. Women we spoke to in these homes talked of their insecurity intensified by their single status. Many women, our participants told us, accept men as live-in partners in order to try and promote greater security for themselves and their children. However, our participants talked of these relationships turning abusive with violence being directed at both adult women and their daughters.

> *Yes, (children) who are not theirs. They mistreat those children and those children find that it's very hard for them and some don't have food or clothes and some try to move where they can get enough clothes. Some are big girls like thirteen, fourteen, they decide to move away from home and go somewhere else or to get married.*
>
> *Participant, Kalobeyei*

Most of the women we spoke with had experienced multiple forms of violence. Applying the ecology model to our data we can see the extent to which VAWG permeates each sphere, from household through to community and the district level with threats and rape from the police. Women and girls clearly feel insecure walking around in public alone which could be somewhat alleviated through adequate lighting in public spaces.

Nonpartner violence

Nonpartner violence happens both within the camp and settlement and also outside. Women, especially those from Kalobeyei settlement, reported incidences of violence perpetrated by outsiders at night, where single women are attacked and raped in their houses and their food rations stolen. This fear of being attacked at night drives them to take in male partners. These kinds of relationships are often detrimental to women and their daughters/children as explained earlier.

> *When women are alone, they can be attacked or even killed. It is common for them and someone might come at night and rape her inside and most of the women have been impregnated. Some of them get into relationships but some of them are forced because you are alone in the house and sleeping alone and someone can come and enter your house.*
>
> *But now the issue is whereby you know in Kalobeyei the (walls of) houses are made of this plastic sheet. And of course, we are not also happy enough at night you feel anytime thieves can come. Like now I have noticed in village 3 next to that big road these people we don't know where they come from and they are dressed in army uniforms and also they come with guns to collect*

> *bamba chakula (food rations) from people at night. So at night they don't sleep and during the day time you sleep but at night you fail to sleep. They demand for food and if you fail they rape you and take everything by force.*

Violence also happens during the day when women are walking from Kalobeyei settlement to Kakuma and vice versa. This kind of violence became so rampant along this specific route that it necessitated a police post to be established midway to improve the situation. Water points have also been associated with violence especially in Kalobeyei, where water is rationed and accessed only on certain days and time.

Distribution centres are seen by some as unsafe. For some women the distribution centres represent insecurity with reports of abuse. In particular a number of women recounted being pushed over and stamped on whilst going to collect sanitary towels and food rations. The food distribution centres are further perceived as insecure as there is rampant theft as women try to manage carrying their rations alone due to the strict rules of a single person per household entering the distribution centre. The women argued in the focus group discussions that there are always people hanging around to grab anything that is left unattended. This is mainly in Kakuma where actual food is distributed. In Kalobeyei the rations come in terms of cash on a sim-card (bamba chakula) that the refugees can redeem for food items of their choice. In these cases, the women are targeted after getting home from the food store with the food.

> *Most of the problems occur in food distribution centres or firewood distribution centers where all refugees go and get their food or the firewood and here most of the time women and girls suffer because they are physically weaker than men and sometimes you are pushed and even myself I was pushed several times and fell down and people walked on me. And particularly when they are giving out the sanitary pads because only women receive but its congested and it's not organized so there needs to be intervention and instead of going to the centre to get the sanitary towels or pads if they can give us money instead of going to the centre if somebody can send me Mpesa or give me cash so that I can go and get these items from the market or nearest shop it would be good instead of receiving them from the centre if somebody can give me cash then I can buy.*
>
> *Participant, Kakuma Camp*

Some of the women reported that because of safety issues it was their husbands/partners who went to the collection centre to collect the rations.

Certain cultural practices like bride price continue to be observed post migration. Bride price was mentioned by a few women as a trigger for the abuse of girls. This is outlined clearly by the following passage.

> *If I stand as a mother and say no you cannot force my daughter to get married by the old man we will have a disagreement between me and my husband, and then we will have conflict and fight. Because I got married using dowry, when we go back to the history and my daughter with that person who has no wealth*

> *and he is just a student and so the father, who is my husband, has to force my girl to get married to that person who has wealth so he can get dowry.*
>
> *Participant, Kakuma Camp*

The risks associated with collecting firewood

Nineteen out of the twenty women in the study cited rape and sexual assault whilst collecting firewood as one of their greatest fears. Some choose not to cook at all rather than expose themselves to the possibility of this kind of abuse.

Our participants claimed that the firewood supplied to the refugees does not meet more than 10% of their cooking fuel needs. The supply is also irregular and sometimes happens once every three months. According to a majority of the women, what is supplied is enough to cover only two weeks of the three months.

This situation leaves the refugees and especially the women, whose responsibility in many cultures it is to fetch firewood and cook in many cultures, to their own devices. They are discouraged from going to the bush outside the refugee camp and settlement to collect firewood by UNHCR and agencies working within the camps, and whilst they have alternative ways of finding fuel many feel collecting additional wood is the most effective. All women in the study attested to trading part of their food rations for firewood at some point every month, but there is a limit to how much food they can exchange for firewood and still have enough left for their families. In the first instance, the food rations are not adequate to cover a whole month according to the women (and especially for the smaller families)[2] and to have to split this out to exchange for firewood in order to be able to cook the remaining food puts them in a difficult situation. It is when they cannot exchange any more food for firewood that they result to collecting from the bush thereby exposing themselves to the violence that lurks therein.

According to the women in the study, the host community feels strongly that all refugee needs should be provided for 100% by UNHCR and stepping out of the camps for firewood is tantamount to trespassing on their land and stealing their natural resources. It is also important to note that the host community uses the firewood and charcoal as a source of livelihood by selling it to the refugees and therefore any direct access by the refugees undermines this and deprives host community members of their livelihood. Conflict and violence can then emanate from the tensions around maintaining this market, which leaves women vulnerable given it is their role to collect the firewood.

For some women – and especially those with young children – the option of not going out into the bush to collect firewood is impossible. If they do not go and collect fire wood their children will go without food and this scenario is worse than the risk of violence. Swapping food for fuel is also not an option as it would result in much smaller meals for their children. Faced with such an impossible situation women organise in groups of not less than five and sometimes up to twenty women and venture into the bush. Some days they are lucky and come back with firewood, other days they are chased out and have to relinquish whatever they have collected in the process of running away or on demand by their attackers. But these women see no other way out of their quagmire and say that they shall continue going out to the bush despite the eminent risks until such a time they are provided with enough cooking energy.

Violence that is meted out during firewood collection includes rape, sexual assault and physical assault. Rape whilst collecting firewood is common. All the women interviewed had either personal accounts of being raped whilst collecting firewood or knew of women who had been.

Women also reported having the firewood they had collected being seized from them and being chased away in the midst of verbal abuse. The fear some women have experienced from being chased by armed young men with intent to harm them has had some of the women vow never to go to collect firewood in the bush again. However, the sexual and physical violence experienced by those caught while in the process of collecting firewood or while running away, made more women stop going to the bushes even when the alternative is that their families go without food for two days as seen in the following quotes. In extreme cases, women/girls have been raped and murdered when caught collecting firewood. The culprits were said to be host community members though some women indicated that there are criminal elements among the refugee community that are capable such acts.

One of the most shocking accounts is given here:

> *I went on 25th of June, so we went to the forest a group of five women and three girls around 10:00am and when we got to the forest we bumped into the Turkanas carrying guns and pangas and they were asking us who authorized us to enter their property and they told us that land doesn't belong to the refugees, that land belonged to Turkanas and not refugees. So they asked us to lie down and we lay down.*
>
> *They had guns and pangas and they beat us and sexually molested us and took one of the girls and raped her, that girl was 14 years. They took one girl. And then we took her to the hospital and she was admitted for a week at Kakuma.*
>
> *I stopped going to collect firewood in the forest after that raping incident and then after that incident another woman also went to collect firewood in the forest and she met a Turkana man and she was able to escape screaming and running back to the camp. But she recognized that it was a Turkana man and that man was seen passing by the camp and we know that it's a Turkana man and even that Turkana was even telling that woman that one day he was going to beat her up.*
>
> *Participant, Kalobeyei*

Participants were asked what alternatives there are to cooking energy in the camps. All the women in the study use firewood as the main source of cooking energy, with charcoal coming in a distant second. The women we interviewed claimed this was the case for all women. No other form of energy was said to be viable within the camps according to the women, though some were aware of alternatives such as gas (LPG), solar cookers and biogas. Some argued that certain kinds of food rations distributed (maize and beans) cannot be cooked through any other means but firewood or charcoal. Not many women were enthusiastic about alternatives that would require them to make monetary contributions, arguing that they have no access to cash within the camps. Refugees do not receive any cash payment since their food rations are provided through mobile money vouchers (bamba chakula) that they can redeem in specific shops for food items of their choice. Those in Kalobeyei settlement receive a flat rate voucher of Ksh 1,400 (£ 10.40) per person per month and those in Kakuma receive actual food rations with a supplement of Ksh 500.00 (£3.70) per month for single person families and Ksh 300 (£2.20) per person to cover food items not included in package.

> *That is both good and bad, bad because if I don't have money to pay and then I will ruin my relationship with the person providing that new cooking method and they might not help me in future. If the instalments are monthly and I don't have any money, I feel like I will be adding more stress to myself and when the payment is due then I will have sleepless nights so that will be a challenge to me but if I have a job as a woman and am sure of being paid at the end of the month and I will budget for that money, and if I can be able to set aside some money to pay for the improved cooking method I would use it.*
> *Participant Kakuma*

In the event that women have sufficient cooking energy that they do not have to stress over what and how to cook, they indicated that they would utilise their time in various ways. They would enrol into vocational centres to learn a skill, attend regular adult education classes and take better care of their children and houses. At the same time, they said they would take time to relax and stress less over whether they were going to provide a meal for their families or not. Above all, for those that currently go to the forests to collect firewood, the anxiety and fear brought by the possibility of being a victim of violence would come to an end.

Lighting within the camp and settlement

Poor lighting in the home and within the camp was mentioned a significant number of times as a trigger for feelings of insecurity and actual violence. Whilst some households have solar lamps given to school children by the United Nations Children's Fund (UNICEF) and other partners, this is insufficient to keep the home lit during the night.

Refugees are not provided with any lighting in the houses, and they have to work this out by themselves. The sources given by the women for household lighting included kerosene lanterns, solar spotlights, mobile phone torches and solar-powered lanterns, while others indicated they have no source of light. However, some of the more established refugees have installed generator powered electricity that they sell to other refugees. Only one participant in our study was able to pay for this kind of light. A number of women who reported using solar lanterns said that this had been made possible because some of their children in grades seven and eight had received these at school to enable them study in the evenings. Lighting at household level is therefore a challenge for most of the refugee families.

Kakuma camp seems to have better public lighting compared to Kalobeyei settlement probably because the camp has been in existence for a longer period of time. However even the lighting in Kakuma is not sufficient as women indicated that only parts of the access roads have security lighting provided by various humanitarian agencies over time. Study participants from Kakuma and Kalobeyei reported that areas without public lighting are more prone to violence at night. Kalobeyei has three villages and only one has security lighting albeit insufficient. The other two villages have no public lighting at night, a situation that criminals have taken advantage of to terrorise residents at night. All women from Kalobeyei knew that village three and two were the most insecure based on the insecurity incidences reported form there. Attacks by unknown people are common at night, with theft of food rations being the highest reported driver of such attacks. These kinds of attacks seem to target female-headed households more than other households, with rape and assault being reported besides stealing of food. Women participants indicated that public security lighting would help a great deal since there were minimal such incidents in village one which is better lit.

> *Yes, when there is the light in my area and my block it will be very safe and reduce violence and when people try to come it would be easy for them to be identified and people who need to attack in the area could be identified because of the good lighting so it would be good when we have good lighting.*
>
> *Participant, Kakuma Camp*

Mainstreaming a VAWG lens

Both the global literature and the data we have presented highlights the extreme vulnerabilities endured by women and girls in refugee settings. Whilst improved energy will not remove them all, we have argued for the benefits of applying greater sensitivity to VAWG in energy programming as a means of supporting a reduction in insecurities. Furthermore, the Call to Action on Protection from Gender-Based Violence in Emergencies (2015) highlights that some risks faced by women and girls may be heightened when humanitarian practitioners overlook women's strength and agency, or when they do not work with local women's organisations and female leaders. It notes that VAWG prevention and risk mitigation efforts should be linked with gender equality work to address existing gender discrimination. In the case of energy programmes this could involve understanding the dynamics of access to energy in the household and community or working with women's groups to identify opportunities for income generation activities as a result of energy access.

Evidence suggests that more humanitarian energy programmes than ever are now explicitly employing a gender and energy lens. The Safe Access to Fuel and Energy (SAFE) initiative was originally established as a protection activity "to reduce exposure to violence, contribute to the protection of and ease the burden on those populations collecting wood in humanitarian settings worldwide, through solutions which will promote safe access to appropriate energy and reduce environmental impacts while ensuring accountability". The VAWG Helpdesk Research Report by Parke and Fraser (2015) outlines strategies that have been employed to improve safety, and includes a brief summary of guidance materials, handbooks and tools used in field operations where access to firewood and alternative energy is a protection concern for women and girls. Listo (2018) agrees with Park and Fraser that the root causes of VAWG are complex and cannot be addressed solely by energy products, particularly in complex and traumatised social settings.

To embed greater sensitivity and encourage a nuanced understanding of the gendered social norms in any given context, the ecological model is a useful and practical tool. The various dimensions of the ecology model could be used by energy programme designers and implementors to steer questions specific to the goals of programmes and to identify the individual, community and social-level factors that drive violence. For example, Table 8.1 offers example questions

Table 8.1 Pertinent questions for energy programmers in relation to the ecology model

Social Ecology Level	*Meta-Question*	*Factors to think about for Operationalisation*
Socio-cultural	In what ways are energy needs and the associated activities gendered?	Can these views be seen to increase the vulnerabilities to violence of women and girls? E.g., if firewood collection became a male pursuit would they experience the same level of risk to violence? And in turn will reduced risk result in changing mindsets?
Community	What community mechanisms exist to mitigate VAWG or offer security and protection to victims? E.g., are there reporting mechanisms following fuel collection rape? And to what extent might this provision cover vulnerabilities linked to energy generations/consumption.	How effective are the community mechanisms perceived to be and could they be built upon by the programme? Shaped specifically to energy needs?
Household	What dynamics exist at household level that may support or perpetuate VAWG?	Are there certain intra-/inter-household behaviours that need to be challenged by the programme? Is there opportunity to use the programme to tap into certain change dynamics (e.g., is there evidence that young educated women challenge their parents about the use of violence)? Can they be supported by the programme?
Individual	What room is there for individuals to challenge and change social norms surrounding VAWG? E.g., do women recognise links between vulnerability to violence and energy-related activities?	Can individuals who may be in the minority but wish to see VAWG end tap into wider networks and support structures to mobilise the change they want to see? If the answer is yes can the programme build on them? If it is no can the programme in fact build them?

Source: Author's own.

that energy programmers might ask and in relation to each of the spheres of the ecology model.

Conclusion

In reviewing the evidence, it is clear that programmes intending to have the impact of reducing VAWG must contain multiple components. Reducing risk will only be effective if it goes hand in hand with social-norm change interventions targeting harmful male behaviours. There are multiple factors that influence GBV in humanitarian settings. There is some evidence to suggest that reducing the trips for firewood collection outside camps reduces exposure to VAWG risks and that additional lighting may make women and girls feel safer. However, such projects only limit vulnerability in one situation, and there is no evidence that they can reduce the actual level of VAWG incidence or that feeling safer is a reflection of reduced VAWG.

VAWG cannot be ended through energy projects alone. Programmes to end VAWG are complex and require a focus on long-term mindset change. That said, the application of a VAWG-sensitive lens in energy programming could well serve to promote a greater sense of control and confidence in women and girls who no longer fear violence.

Notes

1 Another study identified the links in urban and stable settings.
2 Food rations are calculated based on the number of persons in the household and therefore the bigger the family size, the higher the ration. Large families therefore find that 'economies of scale' work better for them than small families of two or three.

Bibliography

Abdelnour, S. A. (2010), 'Fuel-efficient Stoves for Darfur: The Social Construction of Subsistence Marketplaces in Post-conflict Settings', *Journal of Business Research*, 63 (6), 617–629.

Bradley, T. (2010), 'Religion as a Bridge between Theory and Practice in Work on Violence against Women in Rajasthan', *Journal of Gender Studies*, 19 (4), 361–375.

Bradley, T., Gruber, J. and Conroy, K. (2016), 'VAWG Mainstreaming and Framework for Action', *IMC Worldwide*, University of Portsmouth.

CASA Consulting (2001), 'Evaluation of the Dadaab Firewood Project, Kenya. Montreal: CASA Consulting', *CASA Consulting*, Montreal.

Corbyn, D. and Vianello, M. (2018), 'Prices, Products and Priorities: Meeting Refugees' Energy Needs in Burkina Faso and Kenya', Research Paper for the Moving Energy Initiative, *Royal Institute of International Affairs*, London. https://mei.chathamhouse.org/prices-products-priorities.

Danish Refugee Council (2012), 'A Sexual and Gender-Based Violence Rapid Assessment: Doro Refugee Camp', Upper Nile State, South Sudan. https://www.refworld.org/docid/52401d804.html.

DFID CHASE (2012), 'A Theory of Change for Tackling Violence against Women and Girls', CHASE Guidance Note Series, Guidance Note 1. *DFID Practice Paper*.

DFID (2014), 'Review of Non-food Items that Meet the Needs of Women and Girls', June 2014.

Duvvury, N. and Scriver, S. (2014), 'From Domestic Violence to Gender Based Violence: Connecting the dots in Viet Nam', A UN Discussion Paper, *United Nations Viet Nam*, Ha Noi.

Fulu, E. et al. (2013), 'Why Do Some Men Use Violence against Women and How Can We Prevent it? Quantitative Findings from the United Nations Multi-Country Study on Men and Violence in Asia and the Pacific', *UNDP, UNFPA, UN Women and UNV*, Bangkok.

GACC (2016), 'Gender-Based Violence in Humanitarian Settings: Cookstoves and Fuels', *Global Alliance for Clean Cookstoves*, Washington, DC.

Gunning, R. (2014), 'The Current State of Sustainable Energy Provision for Displaced Populations: An Analysis, Research Paper', *Chatham House*, London.

Heise, L. (1998), 'Violence against Women: An Integrated Ecology Framework', *Violence Against Women*, 4 (3), 268–690.

Heise, L. (2011), 'What Works to Prevent Partner Violence? An Evidence Overview'. http://strive.lshtm.ac.uk/resources/what-works-prevent-partner-violence-evidence-overview.

IASC (2015), 'Guidelines for Integrating Gender-Based Violence Interventions in Humanitarian Action: Reducing Risk, Promoting Resilience and Aiding Recovery'. www.gbvguidelines.org.

IOM (2011), 'MRF Nairobi Newsletter', 5 (11). www.iom.int/jahia/webdav/shared/ shared/mainsite/activities/countries/docs/kenya/MRF-Nairobi-Newsletter-May-2011.pdf.

IRC (2014), 'Lighting the Way: The Role of Handheld Solar Lamps in Improving Women's and Girl's Perceptions of Safety in Two Camps for Internally Displaced People in Haiti', *IRC*, New York.

IRC (2015), 'Feasibility and Acceptability of Gender-based Violence Screening: Primary Health Facilities in Humanitarian Settings: Findings from Implementation Among Refugees in Dadaab, Kenya', September. http://gbvresponders.org/wp-content/uploads/2015/09/GBVScreening.pdf.

Jewkes, R., McLean Hilker, L., Khan, S., Fulu, E., Busiello, F. and Fraser, E. (2015), 'What Works to Prevent Violence against Women and Girls: Evidence Reviews', Paper 3: Response Mechanisms to Prevent Violence against Women and Girls. www.whatworks.co.za.

Kelly, L. (1988), 'Surviving Sexual Violence', *Polity Press*, Cambridge, UK.

Krug, E. G., Dahlberg, L. L., Mercy, J. A., Zwi, A. B. and Lozano, R. (eds). (2002), 'World Report on Violence and Health', *World Health Organization*, Geneva.

Lahn, G. and Grafham, O. (2015), 'Heat, Light and Power for Refugees Saving Lives, Reducing Costs', *Chatham House*, London. www.chathamhouse.org/sites/files/chatham house/publications/research/20151117Heat LightPowerRefugeesMEILahnGrafham.pdf

Le Masson, V., Benoudji, C., Sotelo Reyes, S. and Bernard, G. (2018), 'Violence Against Women and Girls and Resilience: Links, Impacts and Perspectives from the Chadian Context', *BRACED*. www.odi.org/sites/odi.org.uk/files/resource-documents/12011.pdf.

Listo, R. (2018), 'Preventing Violence against Women and Girls in Refugee and Displaced Person Camps: Is Energy Access the Solution?' *Energy Research & Social Science*, 44 (2018), 172–177. https://doi.org/10.1016/j.erss.2018.05.012.

Marsh, M. and Purdin, S. (2006), 'Addressing Sexual Violence in Humanitarian Emergencies', *Global Public Health: A Journal for Research, Policy and Practice*, 3 (2), 133–146.

Mercy Corps (2010), 'In Congo, Saving Trees and Lives', 23 April. Mercycorps.org: www.mercycorps.org/ articles/dr-congo/congo-saving-trees-and-lives.

Murphy, M., Arango, D., Hill, A., Contreras, M., MacRae, M. and Ellsberg, M. (2016), 'Evidence Brief: What Works to Prevent and Respond to Violence against Women and Girls in Conflict and Humanitarian Settings?' www.whatworks.co.za/resources/evidence-reviews/item/215-evidence-brief-what-works-to-prevent-respond-to-violence-against-women-and-girls-in-conflict-and-humanitarian-settings.

Okello, S. (2016), 'Research Paper Edited by Stephen Okello (GVEP International) Energy, The Energy Situation in the Dadaab Refugee Camps, Kenya', *Chatham House*, London.

Oxford Policy Management (2014), 'Violence against Women and Girls Map of DFID Programmes'. www.opml.co.uk/files/Publications/a0089-tackling-violence-against-women-girls/vawg-map-dfid-2014.pdf?noredirect=1.

Parke, A. and Fraser, E. (2015), 'VAWG and Energy in Camp-Based Settings', VAWG Helpdesk Research Report No. 94. *VAWG Helpdesk*, London. http://www.sddirect.org.uk/media/1197/vawg-helpdesk-report-94-camp-based-settings-vawg-and-energy.pdf.

Perkins, S. (2015), 'Lighting Impacts SGBV but not in the Way you Think'. http://innovation.unhcr.org/light-impacts-sgbv-but-not-in-the-way-youthink/.

Serrato, B. C. (2014), 'Refugee Perceptions Study Zaatari Camp and Host Communities in Jordan', Oxfam.

Spangaro, J. A. C. (2013), 'What Evidence Exists for Initiatives to Reduce Risk and Incidence of Sexual Violence in Armed Conflict and Other Humanitarian Crises? A Systematic Review', *PLoS One*, 8 (5), e62.

UN General Assembly (1993), 'Declaration on the Elimination of Violence against Women'. www.un.org/documents/ga/res/48/a48r104.htm.

UNHCR (2012), 'Light Years Ahead: Innovative Technology for Better Refugee Protection', *UNHCR*.

UNHCR (2019), 'Kakuma Camp Population Statistics (28 February 2019)'. https://data2.unhcr.org/en/documents/details/68591.

Vianello, M. (2016), 'The Energy Situation in Goudoubo Refugee Camp, Burkina Faso', *Chatham House*, London.

Welland, A. (2015), 'Sexual Violence against Women and Girls: Is Energy part of the Solution?' Smart Villages website: http://e4sv.org/sexual-violence-women-girls-energy-part-solution/.

WFP (2013), 'WFP SAFE Project in Kenya, Kakuma: Fuel-Efficient Stoves and Gender-Based Violence', *World Food Programme*, Rome.

WHO (2013), 'Global and Regional Estimates of Violence against Women: Prevalence and Health Effects of Intimate Partner Violence and Non-Partner Sexual Violence', *WHO*, Geneva. www.endvawnow.org/uploads/browser/files/who_prevalence_2013.pdf.

WRC (2006), 'Finding Trees in the Desert: Firewood Collection and Alternatives in Darfur', *Women's Refugee Commission*, New York.

WRC (2015), 'Call to Action on Protection from Gender-based Violence in Emergencies Road Map 2016–2020'. www.womensrefugeecommission.org/gbv/resources/1240-call-to-action.

Part 3

Future opportunities and solutions

9 Overcoming the data wall

Harnessing big data to understand the energy needs of off-grid communities and the displaced

Iwona Bisaga and Mansoor Hamayun

Introduction

Innovative technologies and their greater affordability, along with increasingly easier and more affordable access to digital devices, have contributed to the transition into the Age of Big Data – a term coined in early 2000s signifying "the explosion in the quantity and diversity of high frequency digital data" (Global Pulse, 2012). Such data has been embraced by the international community as having the potential to inform global, regional and national development and policymaking. It can complement official data sets built through surveys and other more traditional data collection techniques, adding nuance and depth on human behaviour and experiences (Global Pulse, 2012).

This chapter will specifically focus on how big data may be utilised in addressing the energy access gap by providing new and detailed insights into off-grid and displaced communities – and contribute to filling in one of the critical gaps which is the knowledge of the consumers, their needs and ability to pay for off-grid energy services. BBOXX – a next-generation utility providing access to clean energy in Africa and Asia – utilises the Internet of Things (IoT) and Machine to Machine (M2M) tools to build a scalable business model for energy provision and collects energy use data to better understand, and thus better serve, its customers (Bisaga et al., 2017). Recently, BBOXX expanded its Kenya operations to the Kakuma refugee camp where energy access remains an unmet need. This chapter presents a case study of BBOXX's work in the camp and shows how in this environment SHSs can bypass traditional grid infrastructure in favour of smart, decentralised solutions.

Data received from these systems proves that even in a challenging setting such as a refugee camp, consumers can be viable and bankable. It should thus encourage private sector involvement in meeting the energy needs of displaced communities in refugee camps, as opposed to leaving this issue to the donor community to tackle alone, which has historically been the case in the majority of refugee camps around the world. Data, including big data, and knowledge can therefore be harnessed for good: to promote understanding of and investment in this sector, to the benefit of refugee communities and their economies. By doing so, it can contribute to the global data agenda and to the capturing of development data for the global good (IBRD and WB, 2018).

Big data and its challenges

There is no one universally accepted definition of big data. The term can be traced back to the mid-1990s, when scientists began to use it to refer to the handling and analysis of massive data sets with three key traits: volume, velocity and variety (Kitchin and McArdle, 2016). It was popularised in the early 2010s and is now applied more generally to the collection and use of extremely large data sets, which are generated continuously, are scalable and can be analysed using technology in order to draw out detailed patterns and trends, in particular related to human interactions and behaviours (Gandomi and Haider, 2015). Big data therefore goes beyond simple data sets and has changed the way data is captured, stored and crucially analysed. Hashem et al. (2015) assert that the amount of data continues to increase at an exponential rate but not without challenges, as it requires large computational infrastructures to enable its processing and analysis.

Furthermore, as big data is mostly generated by the private sector, including big corporate and technology companies with an extensive customer or end-user base, public-private partnerships will be crucial to ensure data is shared and its benefits maximised in places where they can have the biggest impact (UN, 2018).

In humanitarian settings, the United Nations High Commissioner for Refugees (UNHCR) has already been exploring the effectiveness and validity of using Twitter data to better understand the Europe Refugee Emergency crisis (UNHCR Innovation Service, 2017). The aim of this is to create a real-time situation awareness tool to respond to refugee crises more effectively. With big data, one can analyse the increasing amount of digital data from mobile phones, social media and the internet, and derive useful conclusions from these that can help to alleviate major challenges experienced in humanitarian settings.

However, the gathering and use of big data has been subject to numerous ethical considerations, as it often contains personal and sensitive data (WEF, 2012), which can easily be misused or abused if processed improperly. The UN Development Group (2018) have put together a guide on data privacy, ethics and protection for private sector entities who collect and utilise big data as part of their business offerings. Stress is put on fair, ethical and responsible use of such data in order to extract the most value while at the same time causing no harm or human right violations, particularly given that the current regulations and policies on data (including big data) use in low-income countries are fragmented or nonexistent.

Overcoming those challenges requires making them an integral part of a company's growth and development so that big data's potential, ranging from developing better responses to disease outbreaks, to improving understanding of financial markets, to providing insights into political decision making and much more, can be truly realised.

Big data in the off-grid humanitarian setting

In comparison to the vast amounts of data that exists on consumer behaviour in the developed world, there is a dearth of data in Sub-Saharan Africa and across the developing world. This has led many to assert that this shortage of comprehensive data in the region has hindered action by both the donor institutions and the private sector to find long-term energy solutions for underserved communities. This lack of data can be attributed to the absence of resources and key infrastructure that necessitate the capturing, storing and analysing of data. A deficiency in technological and computing infrastructure, as well as unreliable and sometimes nonexistent electrical grids and telecommunications networks, mean that it is extremely hard for local governments alone to fill this data and knowledge gap on their populations (Luna et al., 2014).

Instead, the unelectrified subset of the population in many parts of Africa and the developing world has often been reduced to rough and inaccurate averages on governmental and multinational data sets. Limited to no ways of reaching those communities has led many investors to view the often rural and/or remote, poor and underserved as one monolithic group, not individuals with complex needs and abilities like in the more connected and reachable parts of the population with financial track records. This particularly impacts displaced and marginalised communities such as those living in refugee camps as even less data is available due to the (assumed) temporary nature of the setting and almost certain lack of infrastructure (Nielsen and Santos, 2013). The lack of knowledge about the inhabitants' payment ability, spending habits and wider socio economic ambitions has left them on the side lines – overlooked or deprioritised by both the public and the private sector as too risky and not worthy of investment (Lehne et al., 2016).

One way to tackle this lack of understanding of these segments is to raise understanding and awareness of such consumers among governments, investors and international businesses, through applying the uses of big data available to private companies who already operate in such contexts and have built a customer base there, among them decentralised energy providers. They can bypass the lacking infrastructure and resources faced by local governments and instead harness the power of big data, for example through the analysis of mobile data, connectivity, and the remote monitoring of solar home systems demonstrating power consumption and usage patterns (Bisaga and Parikh, 2018; Tejwani et al., 2014). Similar data from other key off-grid utilities, such as water providers, can be used to the benefit of both the end users and the public sector by enabling better understanding the viability of such customer segments (Varadarajan and Kaul, 2018; Macomber and Sinha, 2011).

One of the key challenges faced by the displaced living in refugee camps is the assumption that they cannot afford electricity (or other services), yet data has consistently shown the opposite to be true. Big data provides a unique insight into the customers by showing how much they spend on electricity, what time of the day they use it, when they pay for the electricity, as well as what appliances they are using: be it a light, a mobile phone, a television or a refrigerator. With this kind of high-resolution data, demonstrating commercial soundness of service provision, and enhanced understanding of the rural poor or the displaced, the private sector can be encouraged to invest more and to play a greater role in working towards plugging the energy access gap (Bisaga et al., 2017). This makes a great case for mobilising further investment into a sector that institutional investors might have previously seen as the responsibility of the public sector and aid agencies. In addition, data on household affordability and energy use behaviours can also enable donor subsidies and money to be directed more effectively, making interventions more targeted and tailored to the needs of different customer segments.

Data is therefore a vital tool to break down previous assumptions and stereotypes of certain socioeconomic groups and pave the way for investors and the private sector at large to start viewing them as viable and reliable consumers whose bankability can be proven by big data – meaning that the data can show they have sufficient cash-flow to pay for services such as energy. In fact, Lehne et al. (2016) have estimated that energy expenditure of displaced people is around $200 per year per household (approximately $17 per month) and includes cooking and lighting fuels, such as fire wood, charcoal, candles or kerosene for lamps. Given the falling prices of innovative business models and solar energy technologies, including SHSs, which can cost as little as $5 per month on a pay-as-you-go basis, thus boosting affordability and payment flexibility, solar energy can be an appropriate solution for Tier 1 – Tier 3[1] level of access and has already been utilised in refugee camps in Bangladesh, Kenya, Jordan, Ethiopia and Burkina Faso (Fuentes et al., 2018). Such distributed, stand alone energy systems additionally address the challenge of restricted access to energy fuels in refugee camps which tend to be located in remote areas with limited freedom of movement of their inhabitants (Fuentes et al., 2018).

Providing a reliable source of electricity can have a ripple effect too: once people have access, further goods and services can also be provided, offering potential for boosting economic development. For example, by providing access to electricity, customers can more easily charge their mobile phones and use them for business or mobile money; televisions and computers offer access to information along with work and educational opportunities, training in new skills, opening a new business, to the benefit of the end users and their local economies (Uhomoibhi and Damasen, 2014).

Using machine learning and harnessing the power of big data

The International Energy Agency (IEA) predicts that approximately 195 million people will gain access to energy through off-grid solutions, predominantly through solar photovoltaics (PV) (IEA, 2017). In addition, mini-grids, a set of electricity generators or storage systems connected to a larger distribution network that supplies solar electricity to a localised community, as opposed to a single solar home system (SHS), will bring access to a further 290 million people (IEA, 2017). The future of energy in rural Africa and for displaced communities often in remote locations will likely consist of a mix of distributed solutions – from off-grid SHSs to mini-grids and much more (Riva et al., 2018; Schwerhoff and Sy, 2018). However, what is clear is that traditional grid electrification will play a less significant role than it historically has.

The reduction in the price of solar electricity and solar energy storage coupled with more efficient appliances has presented the opportunity to expand energy access to poorer communities such as those living in refugee camps, at a far greater scale. According to the International Renewable Energy Agency (IRENA), the costs for power generated by utility-scale solar PV projects in Africa have decreased by 61% since 2012 (IRENA, 2016). If renewable energy innovation continues at this pace, while incorporating new and innovative business models and finance, it is projected that these costs will only continue to fall. The decrease in costs for solar PV and batteries is making the off-grid solar industry an increasingly attractive and cost-effective solution to the developing world's energy access problem, as well as a viable solution for refugee camps. In addition, in 2016, two-fifths of people in Sub-Saharan Africa had mobile phones, while in the majority of these countries, people with mobile phones outnumbered those with electricity (GSMA, 2018). As such, this widespread uptake of mobile phones in the developing world has been pivotal in driving the off-grid energy revolution – and has encouraged two essential conditions to improve access to energy.

First, it has fostered the rapid onset of mobile money. Countries such as Kenya have been leading the way with the likes of the mobile money service M-PESA. To put this in context, already 74% of adult population have mobile money subscriptions in Kenya and the penetration of the mobile money in the country has been on the upward trend (GSMA, 2017). Due to the level of uptake in mobile money in Kenya and East Africa more broadly over the last few years, the global off-grid energy industry and accompanying investment has been focused on this region (Sharma, 2017). Learnings from the region have recently started to be applied in other parts of Sub-Sarahan Africa, including West and Southern Africa. The rising use of mobile money is highly conducive to the off-grid solar energy sector. This is largely because off-grid SHSs operate most effectively on a pay-as-you-go (PAYG) basis.

In the last decade, SHSs providers have adopted various models most suited to the context of their operations (Guardo, 2018). The particularly prevalent one is the PAYG model which allows users a subscription similar to that offered by mobile phone services, which PAYG in the energy sector has been modelled on and on which it heavily relies to facilitate regular customer payments for energy services (Alstone et al., 2015; Nique et al., 2014; Rolffs et al., 2014). The benefits it offers include lower barriers to customer adoption, transaction cost reduction (for the provider) through the digital payment method and credit risk reduction linked to the real-time, remote monitoring and machine learning (Reichert and Trivela, 2015), and through that the harnessing of big data allowing for meaningful analysis. PAYG platforms have proven to work very well for SHSs and can indeed be replicated at a large scale across a greater range of key services and utilities where they were previously not available (Sharma, 2017). PAYG is is best suited to lower-income populations as the customers only pay for what they use, to pay when they turn it on and stop paying when they turn it off.

Second, the increasing use of mobile phones has also generated greater connectivity. This allows companies to collect money digitally while being able to monitor each system for reliability, as well as enables the company to turn off each system in case of nonpayment, thus incentivising payment performance.

Connectivity can therefore widen the net of financial inclusion by bringing more people into the formal financial system and improve users' financial literacy (FIBR, 2017).

Therefore, the onset of mobile technologies and rising levels of connectivity have contributed to the foundations for the provision of off-grid energy for the unelectrified, and often marginalised segments of the population (GSMA, 2018). With this uptake of mobile comes greater opportunities for collecting data collection which can then be analysed to draw out essential insights lacking in traditionally collected socio economic data sets and proving the scale and the potential of the untapped market for off-grid, decentralised solutions.

BBOXX case study – Kakuma refugee camp

About Kakuma refugee camp, Kenya

This section will look at the case study of BBOXX's work with the Moving Energy Initiative (MEI) – in Kakuma refugee camp in Kenya. BBOXX is a next-generation utility providing access to clean, reliable and affordable energy across Africa and Asia. BBOXX's work has piloted how off-grid solar systems specifically can help refugees without access to basic amenities.

Kakuma is a refugee camp in north-western Kenya in the Turkana County, where 185,000 refugees from neighbouring countries, including South Sudan, Somalia, the Democratic Republic of the Congo (DRC) and Burundi, reside. In addition, it is important to note that while this is a temporary refugee camp, many of the camp's population have actually lived there for over ten to fifteen years. Therefore, much of the housing is semi-permanent, as for many individuals living in the camp there is no likely scenario of returning to their home country due to war, political crises, persecution or other human rights abuse risks.

BBOXX's operations in Kakuma

MEI's data shows that households in Kakuma spend close to $50 per annum on dirty, unreliable and often expensive lighting sources and power, with 22% of camp occupants having no means to light their homes after dark (MEI, 2018). Yet 37% of residents declared improving energy access as one of their top priorities (MEI, 2018). Businesses face similar challenges as residents do in their homes.

While refugee camps are seen as challenging environments for running businesses activities, Kakuma has seen a vibrant informal economy develop over the years, currently giving home to over 2,000 businesses, including grocery shops, small general stores, internet cafes, restaurants and beauty salons (IFC, 2018). As such, there is a high potential for SHSs to power a greater number of small businesses and provide tangible economic benefits to business owners and the local economy alike. Big data can be used to measure and provide key insights into the businesses and customers within the camp by demonstrating the high level of demand for such solutions and showing that these consumers are reliable and have the ability and willingness to pay for solar energy with 0% of customers defaulting in the first few months following purchase, beating the average rate of defaults among customers in other parts of Kenya. This represents a significant opportunity for private sector investment in a previously underserved community.

BBOXX's pilot in the camp was conceived to address these gaps and challenges by setting up distribution outlets to supply and sell SHSs in and around Kakuma camp. In addition to providing a reliable source of clean energy to over 100 households and businesses during the pilot phase, starting operations in the camp has generated the creation of jobs through training local entrepreneurs on running shops, system installation, as well as servicing and maintenance of the systems.

As a starting point to set up the Kakuma operations, BBOXX recruited four shop staff, one retail supervisor, one shop manager and two technicians, alongside nine sales agents from across all sections of the camp. BBOXX has carried out inductions and sales training sessions for the agents and staff and has established and set up a BBOXX shop in Kakuma Town. BBOXX also mapped out the camp into specific wards, assigning sales agents along these lines. Ninety percent of the SHSs sold during the pilot were sold to camp occupants for business uses, among which 25% were female-owned. BBOXX has found that these businesses were able to operate more profitably and stay open for longer as a result of gaining access to a reliable source of energy. One-quarter of all units were sold to women – a number only slightly lower than an average female ownership percentage in other areas of Kenya where BBOXX operate.

In addition, BBOXX and Energy 4 Impact arranged consumer awareness activities around the benefits of solar products. For example, enterprises such as solar kiosks and service centres have been set up, where refugees and host communities can become engaged with clean energy. In particular, there has been an emphasis on showing that while clean energy can provide access to electricity and basic services such as lighting and mobile phone charging, it can offer a range of other amenities, including access to television and small business appliances such as shavers for opening a barber shop. Such activities help turn members of the community into advocates of solar energy – spreading awareness and contributing to improving chances of getting community buy-in.

BBOXX's work in Kakuma has also encountered some challenges in operating in the camp. Logistics costs have been approximately 50% times higher than in the case of other areas where BBOXX have shops, largely due to the lack of existing traditional infrastructure and the remoteness of the location. Lacking communication networks also caused some logistical and transportation challenges. Operational costs of the shop in Kakuma have also been approximately 60% higher than that of other shops due to higher staff salaries, as well as low average revenue per user (ARPU) and small portfolios (compared to the overall BBOXX portfolio). To alleviate shop costs, staff numbers have been cut whereas the recent upsell of TVs has helped increase ARPU. Significantly, there were many language barriers to overcome, as the camp is made up of a variety of populations from different ethnicities, backgrounds and countries of origin. Adverse weather conditions in the area have additionally negatively impacted on the operations, increasing the cost and time of delivery of SHSs to customers.

How big data is harnessed in Kakuma

SHSs provided in the Kakuma refugee camp have been sold and are operated on the same terms as SHSs offered elsewhere in Kenya and other countries where BBOXX provide their services. Likewise, the systems operate on the same technological structures as all other SHSs in BBOXX's portfolio, meaning that the same data is collected, the same switch on-off principles apply and the same systems optimisation opportunities through remote monitoring exist.

These BBOXX solar systems are managed and monitored through a cloud-based task management platform called 'Pulse', which allows BBOXX to efficiently manage customers, products and employees spread across remote and off-grid locations in all countries of operations. In particular, it digitalises the entire sales and service management, automating tasks for sales agents, technicians, call centres and the supply chain. Using modern big data analytical tools – commonly known as machine learning – BBOXX is also able to turn the billions of data points into direct actions. This reduces the need of the middle management to perform complex analysis. Instead, the system is able to tell each technician whom to install, whom to call – as well as providing each shopkeeper with information for whom to follow-up on.

The platform provides the kind of big data which enables the understanding of the environment and of the serviced customers, thus dictating actions directed towards system maintenance and performance control. Cloud-based infrastructure is a critical component in embracing big data and providing an accurate picture of systems' performance. Additionally, machine learning applied in the system infrastructure makes predictive analytics possible, meaning that products can be monitored, and potential failure or battery depletion can be detected early on. As a result, the operations team in the camp is able to support customers before the issue becomes a problem by having appropriate tasks schedule for them. In addition, the data can be used to understand what type of customer is most likely to default (i.e., discontinue payments) and the platform can flag this beforehand. In instances of an at-risk customer, field staff can approach him/her in advance to understand their situation and help alleviate it, where possible, by granting additional grace days needed or offering tips on how to make payments more efficiently, particularly as the PAYG model allows a series of very small payments instead of requiring bulk payments every month.

This ability to translate data into concrete actions carries benefits for both the provider and the end user. A data-driven organisation is able to scale in rural areas and in refugee settings more quickly and in a more targeted manner, reaching both households and businesses with energy services. The ability to predict long-term trends ensures improved experience of the customers as well as sustainability and appropriate planning for the business as AI tools can reveal important details and trends on consumers from the gathered data. Among others, assigning a credit score to each customer, the ability to predict which customer is likely to want to upgrade by adding additional appliances, learning about payment trends and putting mechanisms in place to provide information on how much a customer has paid to date, when he/she needs to pay next to avoid switch off, etc. AI tools also have the capacity to connect sales agents with a certain type of customer to ensure that a successful sale can take place, looking at what customer the agents have historically acquired.

The potential of data gathered from the systems that have to date been distributed in Kakuma to provide unique insights into this poorly understood segment of the off-grid population is great. It has already contributed to gaining knowledge of those adopting off-grid solar solutions in the camp, demonstrating entrepreneurial, bankable consumers with a wide range of needs: from household only to, more frequently, business applications which reliable solar energy can help to power. Through the digitisation of the customers' ability to make regular payments, BBOXX has gained improved understanding of customers' capacity and willingness to pay, making it possible to build a digital profile. However, ethical questions on how else to use the data, whom to share it with and how to gain consent from the customers to do so remain. Although the gathered data is so far used exclusively by BBOXX and utilised to improve the services on offer, and thus customer experience, as contended by UNDG (2017), even if depersonalised, data on vulnerable populations such as refugees should be handled with all the necessary care, and its use should ideally be consulted with those whose data is collected or their representatives. Additionally, ethical and fair ways of sharing such data should be sought to ensure it is used to assist with other interventions which may help the camp's residents, while respecting the nature of the data itself. Currently, customers have a relatively low understanding of what data is collected and how it is used. They are informed about the remote monitoring and switch on/off. However, more complex ideas on how the data is used are not disclosed. Such information might not be necessary to share now but might be needed in the future as more tools are built with the data and more stakeholders get access (on an as and when basis). Those questions around ethics of data collection and management apply to all customers serviced by BBOXX and similar providers, not least those in humanitarian contexts.

Projections

This initial trial in the uptake of solar energy services at Kakuma has yielded positive results despite the challenging setting, particularly in terms of the remoteness of the camp and the limitations it poses to the logistics of servicing it. However, the lack of a formal national grid in this particular region of Kenya shows that alternative routes such as off-grid solar systems may be a viable solution for the

Kakuma refugee camp, Kakuma Town and the satellite townships in the area where high demand for energy services has been observed through the regular stock outs of available SHSs. Another key finding has been the increasing demand for other appliances and services, such as televisions, computers and internet services. Options for deploying a range of services beyond solar for electricity alone will have to be explored and tested. Big data already received from solar systems in Kakuma and that which will be collected as work progresses, and more households and businesses are reached can offer the necessary insights to guide the strategy for scaling up existing services and providing additional ones in the manner suited to the context and the needs of the residents and host communities alike.

Findings from BBOXX's work within the Kakuma refugee camp also align with numerous insights from the installation and servicing of over 200,000 SHSs across other countries where BBOXX operate. It is important to note that energy goes beyond just electricity, with data showing this to be true in Kakuma similarly as in other parts of the world. Electricity is increasingly viewed as the entry point to further economic development. Access to lights alone is a step up the energy access staircase but is not enough to overcome energy poverty, which encompasses other needs associated with energy, such as cooking and access to information services. Providing access to electricity is also the gateway to financial inclusion, allowing customers to build a credit profile and take advantage of the mobile money infrastructure. This can have a spillover effect on what other products and services can become available to those customers. Additionally, by delivering access to electricity for a whole range of products such as TVs, computers, radios, phones and refrigerators, various other dimensions of poverty, beyond energy poverty, can be alleviated.

Existing off-grid infrastructure used for energy services can be translated across to other vital services much-needed for refugee communities, creating a deeper relationship with and thus understanding of the customer. Providing electricity for modern cooking and internet access, and white goods such as refrigerators which can be used to store vaccinations, can improve health prospects in refugee camps as well as in rural areas lacking infrastructure. Under traditional infrastructure development such services would not have been possible or would likely have been too expensive to operate for years or decades to come. The same infrastructure, technology and operating systems which have been developed and are utilised in the provision of off-grid solar energy can further be replicated for water pumping and water access points and potentially other basic yet lacking services.

Finally, using data, mobile and pioneering technology, it can be possible to more effectively mobilise the funds required in investment and manpower to make such prospects a reality by no later than the deadline set by the 2030 Agenda.

Conclusion

Big data has a major role to play in energy access of displaced, off-grid communities and has been a critical technological step enabling improved service provision and learnings which might impact on efforts aimed at achieving holistic development in those settings. The challenges facing refugee camps are indeed exacerbated by the lack of formal infrastructure in off-grid regions across Sub-Saharan Africa. Transformative solutions are therefore needed to be able to reach them.

By looking at the case study of the Kakuma refugee camp in Kenya and BBOXX's work in providing solar systems and raising awareness around energy access, this chapter has endeavoured to demonstrate that despite the challenging environment, big data has a pivotal role to play in humanitarian settings. It can provide knowledge around displaced communities and help bring energy services and other utilities in a manner that is affordable and tailored towards the needs. Thus, data and appropriate technology have a place in supporting efforts to unlock full economic potential of various segments of the population across the developing world. Insights can be harnessed to promote a better understanding of the customers, putting under question the pre-conceived notions that off-grid, often rural and remote refugee communities can only be served by humanitarian organisations rather than the private sector with a utility-level service offering. Big data can help boost confidence among private investors that these are viable consumers willing to pay for energy services, as well as other products and services beyond electricity. Combined with AI technology, big data can fill the knowledge gap that currently exists around refugee populations and help drive investment towards this underserved sector. However, as the sector grows and so does the use of big data, questions around ethics and data privacy will have to be given more consideration to avoid potential data abuses. Similarly, ethical guidelines for data sharing to enable their utilisation to further the development agenda will have to be put in place.

Note

1 According to the Multi-Tier Framework developed by ESMAP (2015), which categorises energy access into Tiers from 0 to 5.

Bibliography

Alstone, P., Gershenson, D., Turman-Bryant, N. et al. (2015), 'Off-Grid Power and Connectivity: Pay-As-You-Go Financing and Digital Supply Chains for Pico-Solar', *Lighting Global Market Research Report*. Resource Document. www.lightingglobal.org/wp-content/uploads/2015/05/Off_Grid_Power_and_Connectivity_PAYG_May_2015.pdf (accessed 5 June 2018).

Bisaga, I. and Parikh, P. (2018), 'To Climb or not to Climb? Investigating Energy use Behaviour among Solar Home System Adopters through Energy Ladder and Social Practice Lens', *Energy Research & Social Science*, 44 (2018), 293–303.

Bisaga, I., Puzniak-Holford, N., Grealish, A., Baker-Brian, C. and Parikh, P. (2017), 'Scalable Off-Grid Energy Services Enabled by IoT: A Case Study of BBOXX SMART Solar', *Energy Policy*, 109, 199–207.

Energy Sector Management Assistance Program (ESMAP) (2015), 'Beyond Connections: Energy Access Redefined', Technical Report 008/15, *World Bank*, Washington, DC.

Financial Inclusion on Business Runways (FIBR) (2017), 'Inclusive Digital Ecosystems of the Future', FIBR Project White Paper No. 2, December, 1–35.

Fuentes, M., Vivar, M., Hosein, H., Aguilera, J. and Munoz-Ceron, E. (2018), 'Lessons Learned from the Field Analysis of PV Installations in the Saharawi Refugee Camps After 10 Years of Operation', *Renewable and Sustainable Energy Reviews*, 93 (October), 100–109.

Gandomi, A. and Haider, M. (2015), 'Beyond the Hype: Big Data Concepts, Methods, and Analytics', *International Journal of Information Management*, 35 (2), 137–144.

Global Pulse (2012), 'Big Data for Development: Challenges & Opportunities', *United Nations*, New York.

GSMA (2017), '2017 State of the Industry Report on Mobile Money', Resource Document, *GSMA Mobile Money*. www.gsma.com/mobilefordevelopment/wp-content/uploads/2018/05/GSMA_2017_State_of_the_Industry_Report_on_Mobile_Money_Full_Report.pdf.

GSMA (2018), 'Mobile for Development Utilities. Achieving SDGs 6 and 7: The Promise and Impact of Mobile Technology', Resource Document, *GSMA Mobile for Development Utilities*. www.gsma.com/mobilefordevelopment/wp-content/uploads/2018/01/Achieving-SDGs-6-and-7-The-promise-and-impact-of-mobile-technology.pdf.

Guardo, A. (2018), 'Innovative Business Models to Power Africa: Distributed Renewable Energy Generation and Aggregation Platforms for Peer-to-peer Trading', *RES4MED & RES4AFRICA White Paper*, 1–16.

Hashem, I. A. T., Yaqoob, I., Anuar, N. B., Mokhtar, S., Gani, A. and Khan, S. U. (2015), 'The Rise of "big data" on Cloud Computing: Review and Open Research Issues', *Information Systems*, 47 (January 2015), 98–115.

International Bank for Reconstruction and Development (IBRD) & The World Bank (WB) (2018), 'Data for Development. An Evaluation of world Bank Support for Data and Statistical Capacity', *World Bank*, Washington, DC.

International Energy Agency (IEA) (2017), 'Renewables 2017: Analysis and Forecasts to 2022', *IEA Publishing*, Paris, France.

International Finance Corporation (IFC) (2018), 'Kakuma as a Marketplace. A Consumer and Market Study of a Refugee Camp and Town in Northwest Kenya', *World Bank*, Washington, DC.

International Renewable Energy Agency (IRENA) (2016), 'Solar PV in Africa: Costs and Markets', ISBN 978-92-95111-48-6.

Kitchin, R. and McArdle, G. (2016), 'What Makes Big Data, Big Data? Exploring the Ontological Characteristics of 26 Datasets', *Big Data & Society*. https://doi.org/10.1177/2053951716631130

Lehne, J., Blyth, W., Lahn, G., Bazilian, M. and Grafham, O. (2016), 'Energy Services for Refugees and Displaced People', *Energy Strategy Reviews*, 13–14 (November), 134–146.

Luna, D. R., Mayan, J. C., Garcia, M. J., Almerares, A. A. and Househ, M. (2014), 'Challenges and Potential Solutions for Big Data Implementations in Developing Countries', *Yearbook of Medical Informatics*, 9 (1), 36–41.

Macomber, J. D. and Sinha, M. (2011), 'Sarvajal: Water for All', *Harvard Business School Case*, 211–228, February (Revised September 2013).

Nielsen, B. F. and Santos, A. L. R. (2013), 'Designing for Multiple Stakeholder Interests within the Humanitarian Market: The Case of Off-Grid Energy Devices', *International Journal of Learning and Change*, 7, 49–67.

Nique, M., Opala, K. and Jain, N. (2014), 'The Synergies between Mobile, Energy and Water Access: Africa', Resource Document, *GSMA Mobile Enable Community Services*.

www.gsma.com/mobilefordevelopment/wp-content/uploads/2014/04/MECS_Synergies-between-Mobile-Energy-and-Water-Access_Africa.pdf.

Reichert, P. and Trivela, U. (2015), 'Increasing Energy Access: The Rise of Pay-as-you-go Solar and Innovative Financing Partnerships', *Enterprise Development & Microfinance*, 26 (3), 248–261.

Riva, H., Ahlborg, H., Hartvigsson, E., Pachauri, S. and Colombo, E. (2018), 'Electricity Access and Rural Development. Review of Complex Socioeconomic Dynamics and Causal Diagrams for More Appropriate Modelling', *Energy for Sustainable Development*, 43 (March), 203–223.

Rolffs, P., Byrne, R. and Ockwell, D. (2014), 'Financing Sustainable Energy for All: Pay-as-you-go vs. traditional Solar Finance Approaches in Kenya', STEPS Working Paper 59, *STEPS Centre*, Brighton.

Schwerhoff, G. and Sy, M. (2018), 'Developing Africa's Energy Mix', *Climate Policy*, 18 (5), 108–124.

Sharma, A. (2017), 'Going Greenfield with Utility Pay-as-you-go Models: Enabling Access to Water, Sanitation and Energy in and Beyond East Africa', Resource Document, *GSMA Mobile for Development Utilities*. www.gsma.com/mobilefordevelopment/wp-content/uploads/2017/12/Going-greenfield-with-utility-pay-as-you-go-models-Enabling-access-to-water-sanitation-and-energy-in-and-beyond-East-Africa.pdf (accessed 12 May 2018).

Tejwani, R., Kumar, G. and Solanki, C. (2014), 'Remote Monitoring for Solar Photovoltaic Systems in Rural Application Using GSM Voice Channel', *Energy Procedia*, 57 (1959), 1526–1535.

Uhomoibhi, J. and Damasen, P. (2014), 'Solar Electricity Generation: Issues of Development and Impact on ICT Implementation in Africa', *International Journal for Information and Learning Technology*, 31 (1), 46–62.

United Nations Development Group (2017), 'Data Privacy, Ethics and Protection. Guidance Note on Big Data for Achievement of the 2030 Agenda', *United Nations*, Geneva.

United Nations High Commission for Refugees (UNHCR) Innovation Service (2017), 'Essys From the Edge of Humanitarian Innovation. Year in Review 2017', *United Nations*, Geneva.

Varadarajan, R. and Kaul, R. (2018), 'Doing Well by Doing Good Innovations: Alleviation of Social Problems in Emerging Markets through Corporate Social Innovations', *Journal of Business Research*, 86 (May), 225–233.

World Economic Forum (WEF) (2012), 'Big Data, Big Impact: New Possibilities for International Development', Resource Document. http://www3.weforum.org/docs/WEF_TC_MFS_BigDataBigImpact_Briefing_2012.pdf.

10 Remote sensing technologies and energy applications in refugee camps

Jonathan Daniel Nixon and Elena Gaura

Challenges to energy interventions for electricity, lighting and cooking

The humanitarian sector lacks data on the success of historical energy interventions and the paucity of data is regularly decried by major actors working on humanitarian energy (for example UNHCR, 2014). Wireless sensors and communication technology offer one potential remedy. Even though energy and cooking issues present in camps have been reported as far back as 1995, it is only since renewable energy technologies have become more established and cost competitive, that solutions and innovations based on these technologies have started to become more widespread (Bellanca, 2014).

In recent years there has been a particular need for data on the use of diesel generators, which are extremely prevalent in the humanitarian sector and often expensive, polluting and inefficiently run (Grafham and Lahn, 2018). It is not recommended that diesel gensets are operated below 50% of their rated load to avoid component damage and reduced efficiencies, but evidence suggests that a lack of understanding on diesel gensets in humanitarian settings means they are often over- or under-loaded in camps, with data on fuel consumption, running hours and servicing not being properly monitored or reported (Grafham and Lahn, 2018). As a result, the cost of generating electricity in UN compounds has been estimated to be very high at around 0.60 $/kWh (WFP, 2017), with an estimated $620 million being spent in 2017 on diesel for electricity generation (and associated maintenance) (Grafham and Lahn, 2018). It is thought that over 450 million could have been saved by implementing affordable alternative technologies and well-established better practices (Grafham and Lahn, 2018). A lack of data on seasonal and diurnal load variations can also make it particularly difficult to provide reliable and cost-effective energy services, which could be easily monitored and reported.

But replacing these diesel generators with alternative renewable energy technologies has not always proved easy and past interventions have had varying degrees of success. Guidelines for the use of portable and modular off-grid solar photovoltaic (PV) systems to provide emergency relief are relatively well-established; however larger micro-grid systems have only just started to be utilised, and research on their operating and implementation in camp setting is limited. The IKEA Foundation's Brighter Lives for Refugees Programme (IKEA Foundation) funded the world's first PV power plant, located in Jordan's Azraq refugee camp, in 2017; and the largest refugee camp PV power plant funded by the German government, followed soon after, located in Jordan's Za'atari refugee camp (Hashem, 2017). Ossenbrink et al. (2018) examined why PV systems have not been adopted more widely in camps. They evaluate PV systems for a range of camp settings highlighting how much lower energy costs would be in comparison to diesel and grid electricity. Following consultation with experts and humanitarian agencies, they attribute the lack of PV systems to be a result of a limited awareness on potential cost and emission savings, suggesting that expertise in the humanitarian sector and in the field need to be improved.

There have been numerous instances of difficulties with using PV technology for street lighting, water pumping and solar lanterns. Lahn and Grafham (2015) review a number of energy projects in refugee camps, identifying that a lack of social and cultural understanding and poor maintenance are often major causes of project failures. Streetlights in Kutupalong refugee camp in Bangladesh were found to be poorly maintained and limited uptake was reported for new solar, energy efficient and gas cookers were highlighted on projects in Burkina Faso, Eritrea and Niger respectively (Lahn and Grafham (2015). The majority of PV-based water pumping solutions in Uganda have been found to reduce costs in comparison to diesel gensets or water trucking (Llario, 2017)[1] and solar-genset hybrids have proven popular for water pumping in many humanitarian settings across Sub-Saharan Africa. However, a lack of water level monitoring and theft of PV panels have been frequent problems. A number of solar-based water pumping stations have also stopped working due to a lack of maintenance. Where systems have been working well, an influx of more people to the area has meant that some systems are now undersized. Aste et al. (2017) investigated a number of hybrid electric systems in use in informal refugee settlements in Lebanon. They identified a lack of local skills to deal with technical issues as they arose. They stressed that monitoring is crucial to deal with unexpected operating conditions when making energy interventions in informal settlements.

Biomass systems have focused on using alternative feedstocks directly for cooking purposes (e.g., pellets or briquettes) or converting a waste feedstock (e.g., food or sewage waste) into a biogas, which can be used for cooking or electricity generation. Biogas units have attracted interest as they have the potential to address multiple issues related to waste and agricultural management, sanitation and energy supply. However, biogas systems trialled in camps have had mixed success. In Haiti, a biogas system comprising of five digesters connected to 60 toilets was built by a nongovernmental organisation (NGO) to provide gas to several community kitchens. The biogas project failed for multiple reasons: technical failures (inadequate gas obtained, lack of water for flushing, pipe damage and gas regulation issues), a lack of accountability for maintenance and socio cultural reasons (community kitchens are not commonly used by Haitians and disputes occurred due to the limited gas supply) (Bellanca, 2014). Many of these technical issues could be monitored with sensors to provide early detection and reporting of gas and water flow problems.

Despite these difficulties, there are nonetheless large potential benefits to unlock. Lehne et al. (2016) assert that efficient cook stoves and solar lanterns could save refugees $303 million a year against their current spending. And thinking beyond products towards delivery models and the entire value chain (e.g., when implementing a new stove, the fuel type, size, cost and collection method needs to be considered to avoid potential problems) could reap even greater benefits (Lahn and Grafham, 2015).

Multi-criteria and multi-objective energy decision making in refugee camps

Research on the use of decision-making techniques for the design and management of energy provisions in refugee camps remains almost completely unexplored. But having better data coming in also means that we can increasingly make better decisions about how energy solutions should be structured in the future. Decision-making models can play a significant role in assisting with camp planning and increasing the success of future energy interventions.

There is now a wide range of energy generation and distribution technology options. However, there are numerous social, technical and economic criteria to consider when trying to deploy and operate sustainable energy services within a refugee camp. Moreover, there are many different conflicting objectives (e.g., minimising capital costs, unit energy costs, network losses and carbon emissions; maximising human development, accessibility and system efficiencies). These objectives can vary for different camps, as do many other constraints and variables (e.g., population size, geographical distribution, available finances and meteorological conditions). To manage these types of decision-making problems, decision support systems based on multi-criteria analyses and multi-objective optimisation methods have become popular tools for renewable energy planning (Pohekar and Ramachandran, 2004).

Many energy project planners need to consider decisions with a whole host of surrounding issues and repercussions. This makes projects complex and difficult to manage. Decision makers will typically use their judgement, experience or intuition to make choices rather than using a holistic and structured approach to evaluate fully decision problems; this can result in poor decisions being made or important aspects of a problem not being sufficiently analysed. Saaty (1980) developed a multi-criteria decision-making method known as the Analytical Hierarchy Process (AHP) to provide decision makers with a systematic method of decomposing a complex problem into several more manageable sub-problems. One major advantage of AHP is its ability to encompass both known and unknown data into the decision rationale; where there are gaps in information, expert opinion and judgement can be used. This flexibility is particularly useful as there are often many uncertainties or subjective data in complex decision problems. Applications of AHP originally demonstrated by Saaty included choosing a school for his son and transportation planning. The AHP is now used all over the world in a vast range of applications including energy planning, marketing, information technology, education and policymaking (Vaidya and Kumar, 2006).

The AHP involves three main processes: 1) identifying the decision goal, selection criteria and alternatives, and organising them into a hierarchy; 2) carrying out pairwise comparisons throughout all levels of the hierarchy and checking for consistency and 3) synthesising all the pairwise comparisons to determine each alternative's overall preference (Saaty, 1988). An example AHP framework for selecting an energy technology for a new refugee camp from a finite number of alternatives is shown in Figure 10.1. In this scenario, it would be invaluable to have data obtained from sensors on prior systems in similar contexts for performance, surplus capacity that goes to waste or activities taking place as monitoring at an individual appliance level.

Despite the benefits of systematic and holistic decision-making models, they have not been utilised for refugee camp energy planning. There are, however, a few examples of such tools being used to address other challenges facing refugee camp sites. Since 2011, hundreds of thousands of refugees have arrived in Turkey from Syria; many remain homeless and new camps need to be built. Çetinkaya et al. (2016) used a multi-criteria approach combined with a Geographical Information System (GIS) to assess new potential camp sites. They used a Fuzzy Analytical Hierarchy Process (AHP) to determine priority weightings for a range of site selection criteria and TOPSIS (The Technique for Order of Preference by Similarity to Ideal Solution) to rank the alternatives sites. A site availability score was determined using GIS and other geographical, infrastructure, risk and social criteria were considered. TOPSIS was used to rank the sites based on an ideal solution formed from a combination of all the sites. Fifteen sites were identified as potentially better locations than the current refugee camps in Turkey; however, as with any decision-making study, the results are highly dependent on the chosen criteria and experts' options. Whilst decision-making tools can be based on expert judgement, their reliability is improved with good quality data.

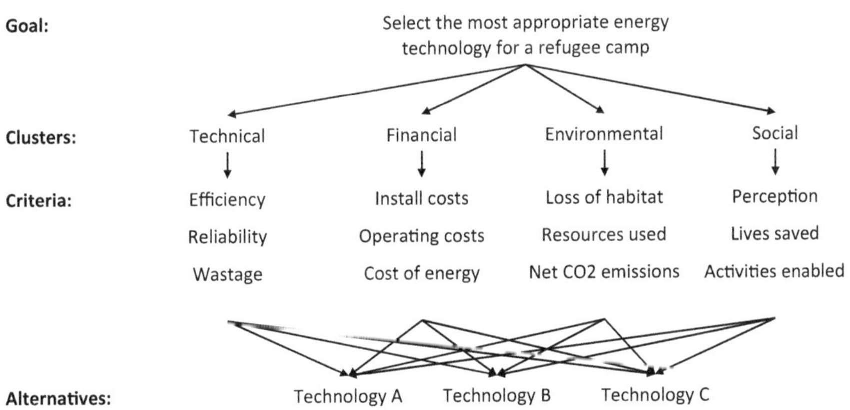

Figure 10.1 Example hierarchy for refugee energy technology selection using AHP

Source: Author's own.

Garfi et al. (2009) used AHP to consider four alternative waste collection strategies for a refugee camp in Saharawi. In addition to technical, financial and social criteria, they took into account environmental impact, human development index and millennium development. Hosseini et al. (2016) used the Integrated Value Model for Sustainable Assessment (MIVES) model (a ranking method which is similar in principal to AHP) for assessing temporary housing units, including water and energy needs. They compared a number of different material and prefabricated components for temporary housing units and suggested that pressed reeds and concrete masonry units were the most sustainable options. Other extensions of AHP have also grown in popularity such as the Analytical Network Process to consider interdependences between criteria (Saaty, 2004) (see Figure 10.2). For example, there is a relation between the cost of a system and technical performance. There are also dependences between social and technical criteria, such as reliability and perception; failure to consider these interdependences can influence the recommendations arising from decision-making studies.

When there is a large number or infinite number of alternatives, multi-criteria decision-making methods can be used to find an optimal solution via an algorithm. For example, optimisation of micro-grids using multi-criteria decision-making techniques is a well-studied area, with researchers investigating both system component and control optimisation. This is due to there being are a large number of different energy generation and storage combinations, which can all be controlled in a variety of different ways. System design decision criteria include capacity of the generating units, type of generating units and financing models. However, there are also many constraints and objectives to consider for system control (e.g., managing peak power demand and equipment loading in order to minimise system inefficiencies and energy costs). Researchers have tackled the problem using numerous optimisation approaches (e.g., linear programming, genetic algorithms

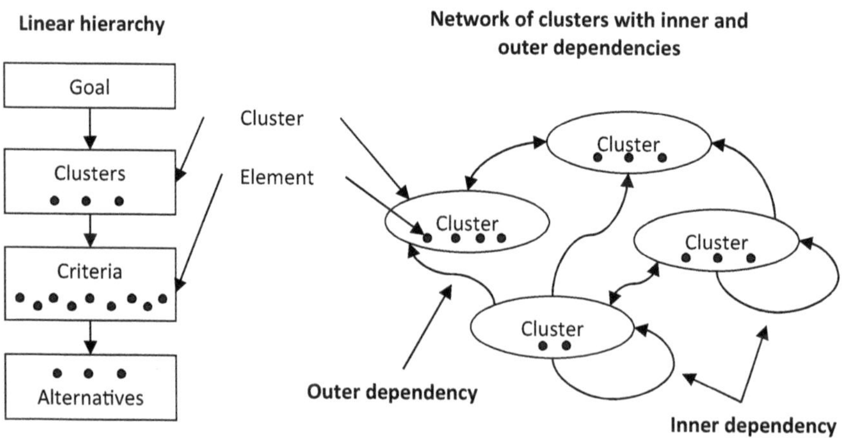

Figure 10.2 Comparison of AHP (left) and ANP (right) frameworks

Source: Author's own.

and particle swarm optimising), considering aspects such as maximising profits and environmental benefits (Tezer et al., 2017; Moghaddam et al., 2012; Sawle et al., 2018; Dawoud et al., 2018). Dufo-López et al. (2016) outlined a micro-grid multi-objective optimisation method for a refugee camp in Tindouf, Algeria. They used a multi-objective evolutionary algorithm for component selection and sizing based on net present cost, human development index and job creation objectives. Using a genetic algorithm, they outline a control strategy for minimising net present cost. A micro-grid solution based on photovoltaic, wind, diesel and battery technologies was outlined, but the authors do not go on to compare their results with the technologies already being used at the case study location, which included diesel generators and PV panels. Therefore, the benefits achieved by the optimisation method remain largely unknown.

Micro-grid control and demand side management strategies

Although most camps do not have access to electricity, there are an increasing number of humanitarian energy interventions, which are prioritising micro-grids. A common approach for micro-grid control is to implement an energy dispatch model, which seeks to minimise instantaneous running costs whilst meeting a total load plus any losses. This is known as an economic dispatch model; however, this approach may not be suitable for refugee camps due to differences in financing arrangements. Demand response programs aim to reduce peak loads by actions being taken by the energy users in response to time varying prices. This would be difficult to implement in a refugee camp, but a simpler request and priority-based approach could be easier for refugees to understand. This would be a form of an incentive-based scheme, designed to incentive users to reduce consumption during peak loads rather than a price-based program, where consumers voluntarily reduce demand based on forward market prices (see Figure 10.3). However, such systems require intelligent interfaces for users to interact with, and engaging with users in this way can be challenging.

To reduce the uncertainty posed by renewable energy inputs into a micro-grid, Aghajani et al. (2017) proposed an incentive-based demand response program

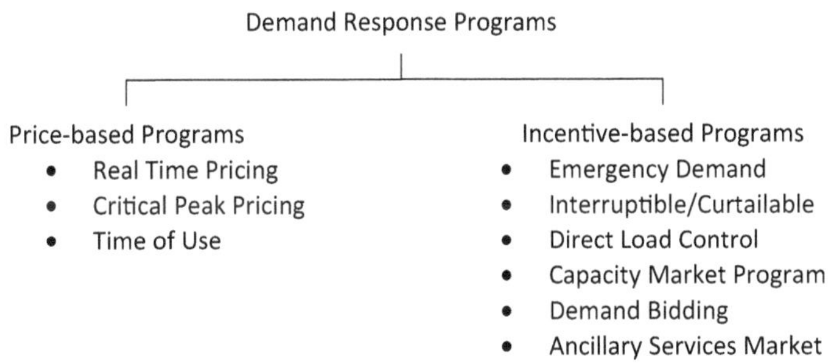

Figure 10.3 Demand response programs can be categorised as price-based or incentive-based, and both involve consumer participation in energy consumption behaviour

Source: Aghajani et al., 2017.

and optimisation method based on probabilistic programming and particle swarm optimisation. The demand response programme objective function was to minimise operational costs and pollution emissions. A number of price package strategies were offered to implement the demand response programme, with lower pricing being offered during low demand periods. Probabilistic programming was used to predict next-day solar and wind generation capabilities. They conclude that if energy users participated in the incentive-based demand response program, costs and emissions could be reduced respectively by 21% and 14%.

Khan et al. (2015) review home energy management systems and demand response program options in various scenarios. They highlight that demand response programs for small residential and commercial areas based on critical peak pricing and load control in the USA have proven very successful. The use of enabling (information communication technology) ICT, sensing and measuring devices is essential to enable smart grid options via a home EMS. A home EMS comprises of a product, which monitors, controls and analyses energy usage within the home, and it functions to reduce energy usage and cost and improve comfort. With the emergence of low-cost and low-power wireless sensors that can be used to control networks and for diagnostics, there is the possibility of them now being used in refugee camp EMS design for the first time to improve automation and intelligence.

The role of sensors for energy planning

Sensors embedded in EMSs are widespread, although a new wave of products and research propose the use of 'per appliance' data to forecast, schedule and control the distribution of energy in renewable based energy delivery systems. Large data sets can now be obtained and stored by energy providers to help them better understand energy usage patterns and emerging behavioural traits. For example, BBOXX, a solar home system provider, is using big data to better understand their systems and predict long-term trends in Kakuma refugee camp (see Chapter 9). Maintenance can be improved by detecting and predicting faults. Payment trends can be evaluated, and operational teams can support their customers more effectively. This information can also be used to inform and devise tools for behaviour change (for example provide user education on effective use of small/limited available energy or help users reduce energy consumption to prevent fuel poverty) or to better plan for/design new energy systems in similar settings. In the context of a refugee camp, quantitative data such as this will help to establish energy patterns and preferences; data that largely remains unknown and is critical to improve energy interventions in camp settings more widely.

Integrating sensors into energy devices, such as streetlights, solar home systems and cook stoves, could enable important information to be gathered that could help to address some of the challenges that have prevented energy interventions in camps from being more successful (Barman et al., 2017; Graham et al., 2014). Where maintenance problems have been encountered, sensors can be used to detect faults and trigger repair services; these sensors can be seen as 'utility meters' that can aid decisions on where assets cease to provide the planned added value or where investment needs to be made due to high failure rates or utilisation. With adequate visualisation and interpretation tools, sensors could warn users of polluting emissions from traditional cookstoves and influence behavioural change (e.g., encouraging the use of an improved cook stove with lower emissions). Table 10.1

Table 10.1 Benefits of sensors and data monitoring of alternative energy options in refugee camps

Technology	*Possible sensors*	*Data to monitor*	*Possible Benefit*
Cook stoves	Thermocouples Emission sensor	Temperature profiles Carbon monoxide and particulate matter emissions	Temperature sensors telling users of efficient cook stoves when they have become too hot and dangerous. Inform households when air quality has become dangerous
Biogas	Flowmeter Thermocouples	Gas yield Digester temperature	Detection of fouling, resulting in the inhibition of microorganisms; improved control of loading rates. Increase in gas yields
Diesel Generator	Bearing sensors	Shaft movement/ vibration and engine conversion efficiencies	Monitor performance/efficiency at low loads Early fault detection and reduced maintenance times and cost. Inform operational change to improve component longevity.
Solar Lanterns	Accelerometer GPS	Movement Location State of charge/ battery health, PV/grid charging rate	To show how lanterns are used by households (e.g., at a standstill in the home during the day or out walking at night, how long are they used, how effectively are they being charged during the day, are they getting wet and damaged during rainy seasons, are they being charged using expensive electricity provided from a diesel genset)
Solar streetlight	Motion and infrared sensors	Mobility and use of space	Better understanding of where and how to operate streetlights in refugee camp settings. Evidence of changes in mobility.
Solar home system	GPS	Operational status Energy availability, usage patterns and preferences.	Theft mitigation/maintenance scheduling Operation change, improved battery life and utilisation efficiencies.
Solar water pump	Ultrasonic sensor	Ground water level Storage water levels	Early water shortage warnings Enable predictive capabilities to improve water management and distribution, thereby mitigating the risk of prolonged periods of water shortages.

Source: Author's own.

summarises some of the alternative sensors to use and data that could be collected on energy intervention in refugee camps and the potential benefits they could enable.

Sensors can be used for improved predictive and automated control (Wu et al., 2010). Imagine streetlamps that can learn patterns of traffic and adapt luminosity and function time, thus enabling energy to be saved and used on other critical services (e.g., health facilities and food refrigeration) and entrepreneurial activities. For water pumping, water levels and quality need to be monitored and, using both forecasted and real-time usage and weather data, water storage and distribution could be managed to avoid long periods of limited water access (Whittle et al., 2013). Similarly, sensor networks are being investigated for improved water irrigation management (Kim et al., 2008). Sensor-based, real-time, decision-making systems in refugee camps will be able to reduce wastage and enable systems to be better balanced. This is critical for camps where fuel, energy and financial resources are limited and evidence-based data on energy usage is unavailable.

One example which brings together these themes is the work undertaken by Coventry University in Rwanda where a smart micro-grid is being deployed to provide power to two nurseries and a sheltered playground. The system is set up to monitor the community's energy needs and patterns of use and show how this changes over time. Another objective of the system is to enable the community to negotiate and share the communal resource. In this instance, an appropriate energy management system (EMS) is needed to control and adjust power flows at an individual appliance level based on priorities and usage quotas. This is particularly needed in a camp setting, where resources are limited and user needs and preferences remain largely unknown, making future system design and control a challenge. Coventry University's work also involves deploying stand alone solar systems and smart streetlights, with a similar focus on characterising energy usage and growth in communal activities. At the time of writing it is too early to say whether the work undertaken in Rwanda will generate data that is truly useful for the sector, but the approach could be instructive for others seeking to contribute to better energy system management of refugee camps.

Conclusion

It is evident that ownership, social acceptance and maintenance are some of the major challenges that need to be addressed to make energy intervention in refugee camps more successful. However, there are a range of tools and technologies which can be utilised to improve the deployment and management of energy interventions in refugee camps.

This chapter has explored the use of multi-criteria, decision-making and optimisation methods, alternative control and distribution strategies and the use of sensors to inform energy system management. Despite the widespread use of multi-criteria decision making for energy planning, it has not yet been taken up for refugee camp energy planning. Refugee camps face multifarious planning problems, which could benefit significantly from holistic systematic decision making, given the wide range of conflicting technical, economic, social and risk criteria and energy technologies solutions that could potentially be used for generation and distribution.

Decision-making tools could play an important role in the future to improve sustainable deployment of energy services in camps. Micro-grid and micro-grid optimisation is another area which has been widely studied, but again it has not been fully explored in the context of refugee camps. Optimal power dispatch and demand side management of hybrid energy solutions remains an exciting area of research for refugee camps as it could significantly reduce energy prices.

Nevertheless, energy usage data still needs to be gathered at camps using sensors, both before and after any intervention, and sensors will play a crucial role in enabling intelligent energy systems to provide cheaper and more reliable services. Sensors will enable systems to be maintained more effectively, reducing downtimes and operating costs, and improve system performance. They will also enable users to use their energy resource more effectively – thereby providing immediate benefits – but they will also capture important data to enable energy technologies to be designed and implemented more effectively to best meet a community's needs in the future.

Note

1 Settlements in Adjumani, Moyo, Yumbe and Arua receiving 90 truck deliveries a month were costing the UNHCR $400,000 per month.

Bibliography

Aghajani, G. R., Shayanfar, H. A. and Shayeghi, H. (2017), 'Demand Side Management in a Smart Micro-Grid in the Presence of Renewable Generation and Demand Response', *Energy*, 126, 622–637.

Amin Hosseini, S. M., De La Fuente, A. and Pons, O. (2016), 'Multi-Criteria Decision-Making Method for Assessing the Sustainability of Post-Disaster Temporary Housing Units Technologies: A Case Study in Bam, 2003', *Sustainable Cities and Society*, 20, 38–51.

Aste, N., Barbieri, J., Berizzi, A., Colombo, E., Del Pero, C., Leonforte, F., Merlo, M. and Riva, F. (2017), 'Innovative Energy Solutions for Improving Food Preservation in Humanitarian Contexts: A Case Study from Informal Refugees Settlements in Lebanon', *Sustainable Energy Technologies and Assessments*, 22, 177–187.

Barman, M., Mahapatra, S., Palit, D. and Chaudhury, M. K. (2017), 'Performance and Impact Evaluation of Solar Home Lighting Systems on the Rural Livelihood in Assam, India', *Energy for Sustainable Development*, 38, 10–20.

Bellanca, R. (2014), 'Sustainable Energy Provision among Displaced Populations: Policy and Practice', *Chatham House, The Royal Institute of International Affairs*.

Çetinkaya, C., Özceylan, E., Erbaş, M. and Kabak, M. (2016), 'GIS-Based Fuzzy MCDA Approach for Siting Refugee Camp: A Case Study for Southeastern Turkey', *International Journal of Disaster Risk Reduction*, 18, 218–231.

Dawoud, S. M., Lin, X. and Okba, M. I. (2018), 'Hybrid Renewable Microgrid Optimization Techniques: A Review', *Renewable and Sustainable Energy Reviews*, 82, 2039–2052.

Dufo-López, R., Cristóbal-Monreal, I. R. and Yusta, J. M. (2016), 'Optimisation of PV-wind-Diesel-Battery Stand-Alone Systems to Minimise Cost and Maximise Human Development Index and Job Creation', *Renewable Energy,* 94, 280–293.

Garfi, M., Tondelli, S. and Bonoli, A. (2009), 'Multi-Criteria Decision Analysis for Waste Management in Saharawi Refugee Camps', *Waste Management,* 29, 2729–2739.

Grafham, O. and Lahn, G. (2018), 'The Cost of Fuelling Humanitarian Aid', *Chatham House*, London.

Graham, E. A., Patange, O., Lukac, M., Singh, L., Kar, A., Rehman, I. H. and Ramanathan, N. (2014), 'Laboratory Demonstration and Field Verification of a Wireless Cookstove Sensing System (WiCS) for Determining Cooking Duration and Fuel Consumption', *Energy for Sustainable Development,* 23, 59–67.

Hashem, M. (2017), 'Jordan's Za'atari Camp Goes Green with New Solar Plant [Home-page of UNHCR]', 29 November. www.unhcr.org/news/latest/2017/11/5a0ab9854/jordans-zaatari-camp-green-new-solar-plant.html (accessed 19 March 2019).

IKEA (2015), 'Brighter Lives for Refugees', 29 November. www.ikeafoundation.org/ (accessed 19 March 2019).

Khan, A. A., Razzaq, S., Khan, A., Khursheed, F. and OWAIS (2015), 'HEMSs and Enabled Demand Response in Electricity Market: An Overview', *Renewable and Sustainable Energy Reviews,* 42, 773–785.

Kim, Y., Evans, R. G. and Iversen, W. M. (2008), 'Remote Sensing and Control of an Irrigation System Using a Distributed Wireless Sensor Network', *IEEE Transactions on Instrumentation and Measurement*, 57 (7), 1379–1387.

Lahn, G. and Grafham, O. (2015), 'Heat, Light and Power for Refugees, Saving Lives, Reducing Costs', *Chatham House, The Royal Institute of International Affairs*.

Lehne, J., Blyth, W., Lahn, G., Bazilian, M. and Grafham, O. (2016), 'Energy Services for Refugees and Displaced People', *Energy Strategy Review*, 13–14, 134–136.

Llario, A. I. (2017), 'Visit Report – Solar and Water Initiative: Refugee Settlements in Northern Uganda – 10th January to 6th February 2017', *International Organization for Migration*.

Moghaddam, A. A., Seifi, A. and Niknam, T. (2012), 'Multi-Operation Management of a Typical Micro-Grids Using Particle Swarm Optimization: A Comparative Study', *Renewable and Sustainable Energy Reviews*, 16, 1268–1281.

Ossenbrink, J., Pizzorni, P. and Plas, T. (2018), 'Solar PV Systems for Refugee Camps: A Quantitative and Qualitative Assessment of Drivers and Barriers', *D-MTEC*, Switzerland.

Pohekar, S. D. and Ramachandran, M. (2004), 'Application of Multi-Criteria Decision Making to Sustainable Energy Planning – A Review', *Renewable and Sustainable Energy Reviews*, 8, 365–381.

Saaty, T. L. (1980), 'The Analytic Hierarchy Process: Planning, Priority Setting, Resource Allocation'. *McGraw-Hill International Book Co*, New York and London.

Saaty, T. L. (1988), 'What Is the Analytic Hierarchy Process?' *Springer*, New York.

Saaty, T. L. (2004), 'Fundamentals of the Analytic Network Process – Dependence and Feedback in Decision-Making with a Single Network', *Journal of Systems Science and Systems Engineering*, 13, 129–157.

Sawle, Y., Gupta, S. C. and Bohre, A. K. (2018), 'Socio-Techno-Economic Design of Hybrid Renewable Energy System Using Optimization Techniques', *Renewable Energy*, 119, 459–472.

Tabaro, N., Gerrard, M. and Quick, D. (2014), 'Safe Access to Fuel and Energy (SAFE): Preliminary Findings from Musasa and Kinama Refugee Camps in Burundi', *Women's Refugee Commission*.

Tezer, T., Yaman, R. and Yaman, G. (2017), 'Evaluation of Approaches Used for Optimization of Stand-Alone Hybrid Renewable Energy Systems', *Renewable and Sustainable Energy Reviews* 73, 840–853.

Unhcr (2014), 'Global Strategy for Safe Access to Fuel and Energy: A UNHCR Strategy 2014–2018', *United Nations High Commissioner for Refugees.*

Vaidya, O. S. and Kumar, S. (2006), 'Analytic Hierarchy Process: An Overview of Applications', *European Journal of Operational Research*, 169, 1–29.

Water Mission (2016), 'Designing Data-Driven Technology with IBM', *Water Mission.*

Whittle, A. J., Allen, M., Preis, A. and Iqbal, M. (2013), 'Sensor Networks for Monitoring and Control of Water Distribution Systems', 6th International Conference on Structural Health Monitoring of Intelligent Infrastructure (SHMII 2013), Hong Kong, 9–11 December 2013.

World Food Programme (2017), 'Energy Efficiency Survey (EES)', *WFP Internal Document.*

Wu, Y., Shi, C., Zhang, X. and Yang, W. (2010), 'Design of New Intelligent Street Light Control System', Control and Automation (ICCA), 2010 8th IEEE International Conference, *IEEE*, Xiamen, China, 9–11 June 2010, 1423–1427.

11 Circular economy in refugee camps

Diego Hakspiel and Johanna Lehne

Introduction

In recent years there has been growing policy and private sector interest in the concept of a 'circular economy' (CE). A circular economy is one in which waste is redefined as a valuable resource to be recovered and repurposed. There has been action on the CE in large multinational companies, in a number of cities in Europe and large legislative frameworks focused on the CE in the European Union (EU) and China. Rwanda, South Africa and Nigeria launched the African Alliance on Circular Economy in November 2017. The CE has also been gaining traction with the international development community with major donors, international organisations and NGOs promoting the CE as a new model for sustainable growth.

So far, however, there have been no studies examining the potential to introduce more circular approaches in humanitarian and emergency situations. This presents a missed opportunity to leverage some of the most effective aspects of CE for post-disaster recovery and humanitarian operations Refugee camps often generate huge amounts of waste which is poorly managed causing health and environmental issues and which could instead offer livelihood opportunities for refugees.

This chapter aims to fill that knowledge gap by identifying opportunities for and challenges to introducing circular practices in refugee camp settings. In contrast to other chapters in this book, it, therefore, does not focus exclusively on 'humanitarian energy' but rather looks at the broader spectrum of resource management, i.e., energy but also water, food and other materials, in refugee camps. Where possible, however, the chapter tries to make the link between more circular approaches and improved energy access for vulnerable populations explicit and highlight the ways in which new humanitarian energy programmes might benefit from looking at their interventions through a CE lens. It draws on a set of existing examples of CE activities and one in depth case study of a current waste-management and energy-generation project by Sanivation in Kakuma refugee camp in Kenya.

What does 'circularity' mean in a humanitarian context?

The basic idea behind a CE is to make the shift from a system in which resources are mined and extracted, turned into products and finally thrown away, towards a system in which the resource loop is 'closed', meaning that waste is reintegrated into the supply chain as a resource (Stahel, 1981).[1] In practical terms this means that when a product reaches the end of its life it is reused, repaired or recycled rather than thrown away. CE also includes shifting to new ways of designing and using products. Sharing a product between people, for instance, may increase the number of times that product is used while decreasing the overall number of products needed. CE, therefore, goes beyond recycling: where it is successfully deployed it not only reduces waste and resource needs but also unlocks additional value from natural resources.

But what does this mean in the context of a refugee camp? Circular approaches can be integrated into many different aspects, from the design and set up of a camp, through to the everyday management and use of resources and finally the dismantling of a camp once it is no longer needed (Figure 11.1). In setting up the camp, for example, specific decisions might be taken on the lay out and location of key services that suit the geographic context and minimise resource use and environmental costs – e.g., not situating accommodation or latrines somewhere that could contaminate fresh water sources. Similarly, tents or shelters could be designed to be durable, reusable and to lower energy needs by harnessing passive heating or cooling systems. On the most basic level a circular approach might simply consist of making sure that solid waste is collected, processed and recycled.

Circular approaches are, therefore, not necessarily novel. There are many existing practices and technologies that are already available and have been employed in humanitarian contexts to help capture added value from resources. The distinction lies in taking a more holistic view to managing resource flows, i.e., rather than thinking about water, sanitation and hygiene (WASH) in isolation, considering how the provision of these services could also facilitate energy access or improve waste management.

How 'circular' are refugee camps?

In some regards refugee camps are more circular than the average context. Sharing living, community and cooking spaces is often the norm, particularly in the immediate period after a crisis. Wastage of food and other resources by those affected by crisis is extremely rare given that many are unable to find enough food, fuel and water. Reuse, refurbishment and adaptation of products and materials that enter the camp are also a fundamental part of life and examples of refugee innovation in these areas are increasingly well-documented (Betts et al., 2015). These activities cover a range of needs from the reuse of electronic scrap to maintain second-hand games consoles in Navikale, Uganda to the adaptation of shelter kits to build shops for refugees in Za'atari in Jordan and the use of broken air conditioning units improve existing cooling systems in Kakuma in Kenya (Betts et al., 2015). Cross et al. (2019) highlight three primary ways in which refugees are implementing more circular processes:

1 Extensive repair systems (for example for wires, batteries and generator systems);
2 Barter exchanges for fuel, batteries, water coolers and other energy objects;
3 Frequent reuse of 'unuseful' energy objects for alternative purposes.

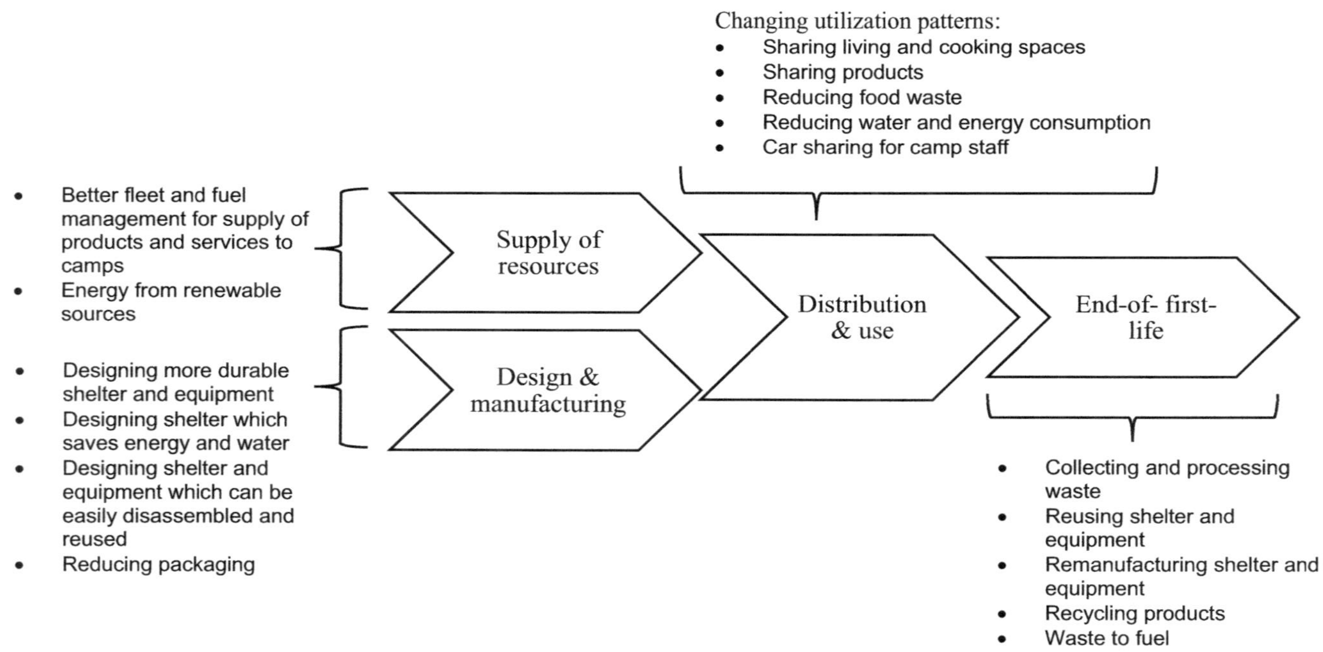

Figure 11.1 Circular economy activities in a refugee camp

Source: Authors' own analysis.

Although this circularity is largely driven out of necessity due to a lack of resources rather than driven by any interest in resource efficiency per se, it is important nonetheless to acknowledge that the ways in which refugees interact with energy services and the way they use resources more broadly are often circular by their very nature. While this aspect of refugee living has previously been somewhat underexplored, ethnographic studies are starting to emerge which show just how central these adaptive processes are to the 'lived' experience of refugees (Cross et al., 2019).

While these examples are positive and show how refugees themselves are already at the forefront of CE approaches in humanitarian communities, refugee camps can also be the least circular settings. Large numbers of people living in often cramped conditions with inadequate facilities can generate huge amounts of unmanaged waste. An environmental impact assessment conducted in 2009 for a proposed refugee camp in Dadaab, Kenya estimated that the new camp would generate 438,000 tonnes of waste that would need to be treated (Price, 2017). To put this into context, Birmingham (UK) generates 500,000 tonnes of waste per year (Birmingham City Council, 2018). Za'atari refugee camp in Jordan generates an extra 480 tonnes of solid waste daily for Mafraq city located 10 kilometres away (Mohammad and Kenneth, 2016). In 2010, the United Nations High Commissioner for Refugees (UNHCR) estimated that around one-third of the 300 refugee camps serving the then 2.4 million displaced people worldwide had inadequate waste disposal and latrine facilities (Coelho, 2010). Beyond poor waste management, refugee camps and humanitarian organisations can be very resource-inefficient in their provision of water and energy services (Grafham and Lahn, 2019). As highlighted in this book, humanitarian agencies spend an estimated 5% of their budgets on fuel (over half a billion dollars globally) and have the potential to make operational savings of over $517 million a year by using currently available, affordable and proven practice and technology changes (Grafham and Lahn, 2019).

Poor resource and waste management can have lasting impacts on sanitary and living conditions for the refugee community and lead to long-term environmental degradation of the surrounding area. In refugee camps in Bangladesh, NGOs are working to remove and decontaminate 3,000 litres of waste a day from hundreds of toilets so that water sources are not contaminated leading to water borne diseases (Aol, 2018). The cholera outbreak in Haiti in 2010 due to poor sanitation practices at a UN peacekeeping camp involved in emergency relief operations in the aftermath of an earthquake highlights the risk of poor waste management (Domonoske, 2016).

Even in instances where waste is collected, it can be disposed of in ways that are damaging to the environment and the health of the refugee community. A study done on waste management in Palestinian refugee camps found that a high proportion is openly burnt – with associated health risks (Cogut, 2016). Moreover, the examples of 'bottom-up humanitarian innovation' discussed earlier may not always be the most effective means of processing waste, especially at scale. Refugees often lack the technology to optimise recycling and repair processes, and refugees picking up waste in camps face health hazards, lack protection and have to deal with the stigma attached to their work (Bhalla, 2018).

What drives this?

One reason why waste management does not tend to be prioritised is that refugee camps are often intended to be temporary solutions to emergency situations. They are often hastily constructed in response to a fast-developing context. Over a period of a few weeks in July 2016, for example, more than 4,000 people crossed the border from South Sudan into northern Uganda daily (Edwards, 2016). Responding to this influx put a massive strain on collection points, and transit and reception centres, as NGOs rushed to build temporary shelters and increase assistance capacities as quickly as possible. In such circumstances, it is unsurprising that waste management takes a backseat.

At the same time the countries in which refugee camps are located often do not want to recognise that these situations might become long-term and therefore discourage the construction of longer-term infrastructure and housing. While it may be expedient to emphasise the temporary nature of a refugee camp from a political standpoint, this can be costly in circumstances where camps turn into permanent living circumstances lasting for decades or more – a trend which is becoming all too common.

In addition, the actions of NGOs and humanitarian organisations themselves can exacerbate the situation. Humanitarian responses can generate waste from packaging of emergency supplies to medical waste from treatment of the camp population. Well-meant cook stove and off-grid energy interventions have generated additional waste in refugee camps where waste management processes for these cooking and solar products were not in place (Magalini et al., 2016).

Opportunities for doing more with less

Introducing sustainability or 'circular' principles to refugee camps does not have to interfere with the concept of providing only 'temporary solutions'. CE design principles such as modularity, for example, can allow for durable shelters to be built which can be easily taken apart, moved and reused elsewhere when they are no longer needed without generating huge amounts of waste.

Moreover, to date refugee management has largely overlooked the potential of waste management and recycling for improving livelihoods in refugee camps. Waste produced in camp can be collected and processed by refugees, used to generate energy and produce products and equipment by refugees for refugees – both reducing resource needs and providing occupation for refugees. A project in Lebanon's Bekaa region teaches refugees to sort waste, recycle and compost organic waste and allows them to earn money collecting recyclable waste in camps (Clasper, 2016). This can have positive social impacts, giving refugees a feeling of self-sufficiency and community.

The self-sufficiency and resilience that 'circular' approaches could deliver are particularly important for refugee camps, which are generally located far from cities and often have to rely on little space and tightly packed living conditions. The more sustainable and resilient these locations are the better. Moreover, better resource management could lower the environmental impacts of camps and the impact on the wider community, addressing social tensions between host and refugee communities.

Some of the countries in which large refugee populations are located have recently introduced stringent policies to improve waste management. Rwanda and Kenya have each imposed total bans on plastic bags in a bid to stem growing waste crises (de Freytas-Tamura, 2017). Nigeria, South Africa and Rwanda are working with the World Economic Forum to develop an Africa-wide alliance to promote the development of a CE.

Existing 'circular' innovations in refugee camps

Although the examples highlighted earlier indicate the ways in which refugee camps are currently not very resource-efficient ways and some of the barriers to improving processes, there already a number of examples of circular initiatives in refugee camps. The types of interventions can be organised into five broad categories: improving waste collection, improving waste processing, waste to energy, waste to equipment/products and processes and better design (Table 11.1).

These examples span many different types of circular approaches – from NGO-led training courses to technology and infrastructure being brought in to upgrade camp operations. Not all of these interventions necessarily always represent the best use of resources. Waste-to-energy projects, for example, have received a lot of funding from international development actors. They offer an easy fix to both energy access issues and to growing volumes of unmanaged waste in camps. However, they are often criticised for dealing with one problem (volumes of waste being dumped) by causing another problem (greenhouse gas emissions and health impacts from burning that waste) and for diverting waste from better uses such as recycling, remanufacturing or reuse.

There are, moreover, few examples of organisations helping to scale up refugee-led circular innovations. The success of many of those innovations relies on the technical expertise of a refugee or group of refugees and their ingenuity in adapting those skills to the everyday problems they face in the camp. In an ideal scenario, these skills would be harnessed and refugees would be more closely involved in the design of camp products and services to improve resource efficiency (Cross et al., 2019). In practice, however, this requires a degree of flexibility and adaptability that is often understandably lacking in organisations involved in camp operations. In Za'atari, in Jordan, organisations have highlighted the challenges of providing services in a camp that has grown semi-organically as Syrian refugees have adapted and transformed spaces and shelters to their own needs (Betts et al., 2015).

Table 11.1 Examples of circular innovations for/in refugee camps

Type	*Description*	*Examples*
1. *Improving collection and separation*	Establishing better processes for collecting waste.	• Oxfam recycling project in Za'atari and Mafraq in Jordan to improve waste separation by households in the refugee camp.[2] • Arcanciel project in the Bekaa region in Lebanon, teaching refugees to collect recyclable waste, which is then sold on to external companies.[3]
2. *Improving processing*	Establishing better recycling, remanufacturing and reuse of waste.	• GIZ has established seven recyclables sorting stations in Jordan to generate employment in the recycling sector for Jordanians and Syrian refugees.[4] • GIZ funded ACTED project encouraging segregation of waste at household level in Azraq camp in Jordan and promote recycling.[5] • Kenya Red Cross Society waste recycling project in Dadaab, Kenya.[6]
3. *Waste to energy*	Using waste to generate energy, e.g., making briquettes out of waste, burning waste as fuel	• EU-funded Oxfam project to produce briquettes from solid waste refugee settlements in northern Uganda.[7] • UNHCR project in Cameroon, teaching Nigerian refugees to make agricultural waste into 'eco briquettes'.[8] • Sanivation project in Kakuma refugee camp in Kenya turns human waste into fuel.
4. *Waste to equipment, products, other processes*	Using waste to build new products or processes	• Refugee Utility Project is teaching refugees how to make air conditioning units out of plastic bottles in northern Jordan.[9] • IVE Engineering's self-powered mobile recycling facility for turning plastic bottles into reusable building blocks.[10] • Ziadat for Recycling working with UNHCR and UNDP to teach refugees and Jordanians to upcycle furniture.[11] • NRC project in Za'atari camp in Jordan to repurpose waste materials into hydroponic pot systems – systems for growing plants without soil.[12] • Tateh's sand-filled plastic bottle houses in Sahrawi refugee camps in Algeria.[13] • Matériaux sans frontiers, a project run by Architecture sans frontières Québec, aims to facilitate the donation and recovery of building materials for humanitarian settings.[14]
5. *Better design*	Designing packaging/products/systems so that they last longer, are more energy-efficient, create less waste and easier to disassemble at end-of-first-life	• Partnership between Edinburgh Direct Aid and Smurfit Kappa. They send products to refugee camps in boxes designed to be converted into stackable stools, storage chests and desks once the products have been delivered.[15] • Ikea's solar-powered flatpack Better Shelter has been designed to last longer than a typical emergency tent.[16] • Blueprint 200 project to rethink the design of camps to encourage more sustainable water management, greener ways of cooking, disposing of waste and generating electricity.[17]

Source: Authors' analysis of various sources.

Finally, many of these interventions face barriers, including access to the requisite technology and finance. To explore these more fully the next section looks at one of these examples in detail.

Waste to value in Kakuma refugee camp

This case study focuses on a project by Sanivation, a Kenyan social enterprise delivering cost-effective sanitation services by transforming human waste into a fuel product, in Kakuma refugee camp. The first part of this case study will provide an overview of Kakuma refugee camp and the sanitation situation in the camp. It will then focus on the sanitation solution delivered by Sanivation and the transformation process from waste to value. It concludes with lessons learnt and potential opportunities available for circular economy practices in refugee camps across the world.

Kakuma refugee camp

Kakuma refugee camp is located in the northwest of Kenya, 120 km away from the South Sudanese border. The camp was built in 1992 to temporarily host approximately 20,000 refugees from Sudan and Ethiopia. By the beginning of 2018, the camp had expanded to host over 185,000 refugees from over eight different nationalities (UNHCR, 2018a). This increase in population has challenged the capacity of humanitarian agencies to deliver aid effectively and forced the humanitarian space to look into innovative solutions.

The goal of UNHCR in long-term refugee camps, such as Kakuma, is to provide at least 85% household latrine coverage (UNHCR, 2018b). According to data gathered by UNHCR in December 2017, household latrine coverage for Kakuma refugee camp stood at only 41% (Sukru, 2017). Low latrine coverage is, in part, the result of the increased influx of South Sudanese refugees since 2014, in combination with a decrease in global funding for long-term refugee camps. In addition, the standard sanitation solution provided in Kakuma, the traditional pit latrine, has increasingly become incompatible with the camp due to the difficult ground conditions the camp faces.

Difficult ground conditions that hamper building new pit latrines include:

- *High water tables:* water supply can get contaminated with faeces in pit latrines;
- *Flood prone areas:* pit latrines overflow when it rains leading to pits collapsing and an increased risk of disease transmission;
- *Rocky soils:* this type of soil increases the difficulty of digging new pits;
- *Lack of space to dig new pits:* decommissioned pit latrines take up valuable space and new refugee arrivals make space increasingly scarce.

In the face of these difficult conditions, how can aid agencies provide sanitation for an increasing population with less funding and less space available? These were the challenges Sanivation aimed to address by transforming waste into a value-added product.

Sanivation: waste to value

Sanivation is a social enterprise based in Kenya. Its mission is to increase access to safe and cost-effective sanitation services in urbanising communities and refugee camps (Sanivation, 2018). The company first piloted its sanitation system in Kakuma in 2014 (Boston Consulting Group, 2015). From 2016 to 2018, the system was implemented at scale to cover 2,500 people (500 households). The system addresses the entire sanitation value chain as shown in Figure 11.2.

Sanivation provides container-based, urine-diverting toilets, biweekly collection of faecal waste and treatment and transformation of faecal waste into an income generating fuel product. The system goes beyond basic sanitation and ensures safely managed sanitation for households in accordance with Sustainable Development Goal (SDG) 6.2.[18]

The first two components in the value chain (toilet provision and bi-weekly collection) aim to address the difficult ground conditions presented in Kakuma. Following a human-centred design approach (IDEO, 2018),[19] Sanivation designed an above-the-ground household toilet that contained faeces safely and could be regularly emptied. By being above the ground, the new toilet avoided pit digging, mitigating the risks of toilets overflowing after rain and of water contamination due to high water tables. The ability to regularly empty the toilet increased the lifespan of the toilet and decreased the need for space to build new ones. Sanivation also trained and employed refugees to safely collect the waste from used toilets every three days. This allowed refugees to take ownership of their own sanitation system and created an income-generating activity within the camp. Overall the new toilet design and collection service addressed all the difficult ground conditions

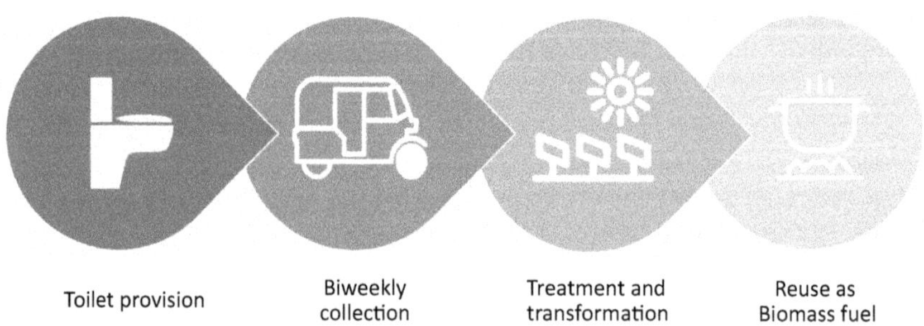

Figure 11.2 Sanivation system overview

Source: Diagram from authors' own analysis.

previously identified. However, Sanivation still had to provide a solution for what to do with all the waste being collected.

Based on data gathered by Sanivation in Kakuma, one household produces on average 1 kg of faeces per day.[20] At a scale of 500 households that translates to 15,000 kg of faeces being collected every month. Sanivation aimed to lower the costs of waste treatment by using resource recovery principles. Following a market-based approach, the company looked into potential market gaps that a faeces-based product could help address. Cooking fuel was identified as a potential market gap, as Lokado, the local NGO in charge of providing cooking fuel to refugees, was only able to provide 10 kg of firewood per refugee every two months. That amount only accounted for one-third of the camp cooking fuel needs. The remaining two-thirds were up to refugees to self-provide, leading to significant deforestation of the area (Owiiti, 2003). If Sanivation was able to transform faeces into cooking fuel that was socially accepted, cost-competitive with firewood and charcoal, and compatible with the existing stove options, they could generate revenue out of faeces, safely dispose of all the waste collected and help reduce deforestation by providing an alternative to the existing cooking fuel options.

The last two components of the value chain (treatment and reuse) aimed to provide an answer for what to do with all the waste collected, while at the same time delivering a more cost-effective sanitation solution. Having identified a market gap for a potential faeces-based fuel product, Sanivation selected charcoal briquettes as the best option for the existing market. Charcoal briquettes were compatible with existing stoves and could be sold at the same price as traditional charcoal. In addition, briquettes allowed faeces not only to be used as a heat source (dry faeces have a heat content value of 17 MJ/kg) but also as a binder for other biomass waste inputs that, combined with faeces, could increase the final briquette production output, and in turn its revenue potential. One such waste input is charcoal dust. It is estimated that between 10% and 20% of wood-based charcoal turns into dust during transportation and storage. In Kakuma, charcoal dust is traditionally thrown away and considered waste without value. Sanivation recognised this additional carbonised bio-waste as an opportunity.

Heating faeces to temperatures above 65°C for a minimum of three hours is one of the most effective means to render faecal waste safe for reuse (Foote et al., 2017). Based on this knowledge, Sanivation designed and deployed the first ever waste treatment plant in Kakuma. The system was fully operated by local refugees and host community members and had the capacity to safely treat over 2 tons of faeces per day. Once treated, faeces could be transformed into charcoal briquettes. Briquettes were produced by mixing treated faeces with charcoal dust and using a roller press to mould the mix into small pillow shaped briquettes compatible with refugee stoves. By carefully controlling the production process and by using faeces as a binder, Sanivation was able to produce briquettes with a calorific value higher than wood, and with lower smoke emissions than traditional charcoal as shown in Table 11.2.

Table 11.2 Fuel comparison between charcoal, wood and Sanivation briquettes

	Charcoal	*Wood*	*Sanivation briquettes*
Calorific value	29 MJ/kg	15 MJ/kg	22 MJ/kg
Burning time	3 Hours	1 Hour	4.5 Hours
Emissions (CO ppm)	118 ppm	NA	82 ppm
Emissions (PM2.5 ppm)	213 ppm	NA	196 ppm

Source: All fuels were tested according to the water boil test (WBT) testing protocols[21] using a Kenya Ceramic Jiko[22] (KCJ). Tests were carried out independently by the University of Nairobi, and the Kenya Industrial Research and Development Institute (KIRDI).

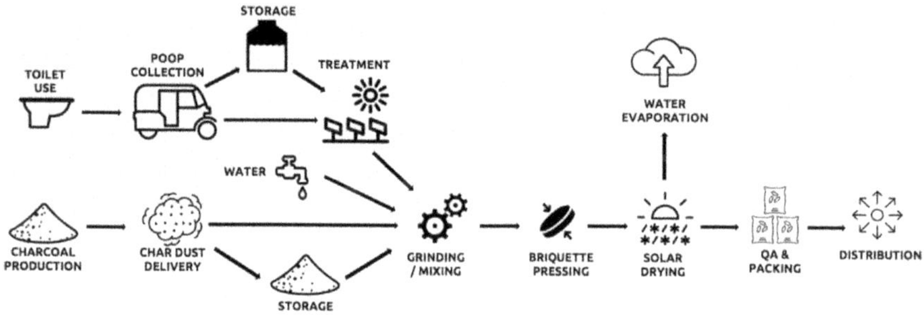

Figure 11.3 Flow chart of Sanivation's system

Source: Diagram from authors' own analysis.

Figure 11.3 presents a flow chart of the entire sanitation system implemented by Sanivation. The flow chart showcases how the two waste inputs described in this case study were transformed into a valuable fuel product.

Cost-effectiveness and comparison

The net costs[23] of Sanivation's system at a scale of 500 households are estimated at USD60 per household per year in comparison to USD65 per household per year for pit latrines (Table 11.3). These costs account for start-up costs and local refugees managing the system themselves but do not include management expenses for the implementing partner, as this information was unavailable at the time of the study. While Sanivation's system has higher upfront costs than pit latrines, it is estimated to be cheaper than pit latrines in the long term due to the revenue generated from the briquette sales. In addition to cost savings the system has significant non-financial benefits that are important to camp administration and local government. By leveraging faecal and biomass waste available in the camp, the system creates significant revenue and has the potential to offer better value for money in the long run for humanitarian agencies (UNHCR and Boston Consulting Group, 2015).[24]

By the end of 2018, Sanivation was able to validate the total costs per household (HH) per year[25] and the briquette revenue potential per household per year.

Table 11.3 Sanivation's system costs vs pit latrines

	Sanivation Approach	*Household pit latrine*
Total cost per HH/year	$250 USD	$65 USD
Briquette revenue per HH/year	$190 USD	–
Net cost per HH/year	$60 USD	$65 USD

Source: Costs presented in this table are based on operational data collected by Sanivation between 2016 and 2018.

Note: Estimated lifespan of pit latrine = two years

The word potential is used to highlight that annual briquette production capacity and price of sale, were both validated at operational level, but the total quantity required to achieve the net costs shown in Table 11.3 had not yet been sold in practice. Additional time was required to demonstrate in practice, that annual briquette production could be sold within one year.

Non-financial benefits of Sanivation's system include:

- *Reduced deforestation:* 88 trees saved per ton of briquettes produced (Baconguis, 2003)
- *Improved health:* toilets contain faecal waste safely and briquettes produce fewer carbon emissions;
- *Livelihood creation:* the system is designed to be locally managed by refugees and the host community.

Lessons learnt and future opportunities

As illustrated earlier, Sanivation was able to address a range of sanitation challenges by transforming human waste into a valuable product. Difficult ground conditions were addressed by safely containing and removing human waste from the community, and the valuable fuel product created at the end of the process supported the need for innovation in a challenging funding context. It is important to note that Sanivation's end goal was never to simply implement a circular economy process but rather to deliver value to its customers (humanitarian agencies and refugees) through a more cost-effective approach. As global humanitarian funding continues to decrease, there is a significant opportunity for private sector and humanitarian partnerships that combine both sectors' expertise to deliver more cost-effective solutions for beneficiaries and to maximise the use of local resources.

There is no one-size-fits-all solution. Addressing the entire sanitation value chain allowed Sanivation to be flexible and adapt to different opportunities and challenges. However, availability of inputs, local markets and social preferences all vary in different contexts. In the case of Kakuma, Sanivation decided to produce charcoal briquettes as it was the most profitable product it could make with the waste materials available, while ensuring there was a readily available market for it. Transforming human waste into fertiliser would have been a much easier process and was initially considered as an option. However, as Kakuma is located in a desert, with little to no agricultural activity due, such a product would have resulted in further waste of resources instead of optimising their use.

Introducing the use of faeces-made briquettes was not easy to sell to humanitarian implementers and project funders, with many raising concerns that briquettes would not be accepted by refugees due to negative cultural norms around human contact with faeces. Sanivation was able to demonstrate that product user acceptance is largely driven by product performance and pricing. Creating a product that significantly outperformed the competition by burning longer and producing less smoke allowed Sanivation to overcome negative cultural norms around faeces-based products. Sanivation also engaged local leaders as product champions, carried out cooking demonstrations, and provided free samples to gain users' trust and overcome cultural preconceptions.

By the end of 2018, Sanivation's project was only processing a small fraction of Kakuma's entire faecal production. With a population of more than 180,000 people, the camp produces on average 13,000,000 kg of faeces a year.[26] At such a scale, the production of cleaner fuel products such as 100% faeces pellets, biogas and electricity could become financially viable, generating higher revenues than charcoal briquettes and allowing the camp to transition entirely into non-wood-based fuel products.

However, a number of factors limit the scale up of Sanivation's project:

- *Access to financing to cover upfront costs:* operational budgets in refugee camps are allocated on an annual basis. Requesting upfront costs for a technology with a five to ten year+ lifespan can be challenging under a traditional humanitarian procurement process. This could be addressed by integrating a variety of existing public-private partnership (PPP) (World Bank, 2017) financing mechanisms into humanitarian procurement, and through private capital coupled with multi-year contracts.
- *In-kind refugee assistance from humanitarian partners* makes it challenging for private sector products to compete against free alternatives such as free firewood provided by humanitarian partners. This is particularly relevant for products where the main value proposition is cheaper pricing or value for money. A solution to this challenge is the implementation of cash-based interventions (CBIs) that allow all products to compete fairly in an open market and give refugees the choice to select the product that best fits with their needs.
- *'Buy-in' from the humanitarian sector* on innovative solutions is usually dependent on validated operational data at significant scale to inform decision making and reduce risk of investment. There is availability of funding for piloting innovative solutions at scales around 50 to 500 households. However, funds become significantly scarcer for projects at 1,000+ households requiring over 1,000,000 USD in financing. Organisations such as the Bill and Melinda Gates Foundation (BMGF), and grant opportunities such as the

'Creating Hope: A Humanitarian Grand Challenge',[27] are alternatives to de-risk upfront investment; however, they are significantly scarcer and far more competitive.

- *The lack of clear guidelines for private-humanitarian partnerships:* there is little clarity on how private and humanitarian partners should work together without having to find loopholes within the existing humanitarian procurement guidelines. Topics such as revenue and patent ownership in humanitarian funded projects remain unclear under the existing guidelines. This lack of clarity can lead to longer timelines for contract signing and can dissuade the private sector from engaging with humanitarian partners. Developing clear partnership frameworks could streamline the process and incentivise more private partners to see the revenue potential in refugee camps.

The challenges experienced in Kakuma refugee camp are not exclusive to this particular setting. Difficult ground conditions, decreasing humanitarian funding, lack of sustainable fuel options, risk of sexual and gender-based violence when collecting firewood and a lack of livelihood opportunities are challenges experienced in refugee camps across the world. As Sanivation continues to scale up, they have already validated system compatibility in Rhino camp (Uganda), Gambella region (Ethiopia), and Nyarugusu camp (Tanzania). All of these locations present opportunities for the implementation of cost-effective circular economy approaches that could help improve the lives of refugees.

Conclusion

More circular approaches are possible in humanitarian settings. The examples cited earlier of existing circular practices, and of refugees engaged in innovative activities to use the resources at hand in the most efficient way possible, provide excellent entry points for scaling up circular activities. The Sanivation project highlights the many benefits from transforming human waste into a valuable product. But also emphasises the political, financial and technical constraints that are likely to present challenges to such projects. Where these barriers can be overcome, shifting towards a more 'circular' refugee camp could generate environmental, social and financial benefits for the refugees living in the camps, the organisations running them and the local host community.

Notes

1 The term 'closed loop' was used as early as 1976 by Walter Stahel, considered one of the key industrial ecology thinkers. Stahel, W. (1981), 'Jobs for Tomorrow: The Potential for Substituting Manpower for Energy', *Vantage Press*, New York.

2 Cuthbert, O. (2016), 'Broken Glass and Needles: The Waste Pickers Scraping a Living at Jordan's Landfills', *The Guardian*, 27 August 2016, www.theguardian.com/sustainable-business/2016/aug/27/jordan-waste-pickers-syria-refugees-recycling (accessed 14 May 2018).

3 *Sustainia* (2017), 'Solid Waste Management in Refugee Camps', http://explorer.sustainia.me/solutions/solid-waste-management-in-refugee-camps (accessed 14 May 2018).
4 Cuthbert, O. (2016), 'Broken Glass and Needles: The Waste Pickers Scraping a Living at Jordan's Landfills', *The Guardian*, 27 August 2016, www.theguardian.com/sustainable-business/2016/aug/27/jordan-waste-pickers-syria-refugees-recycling (accessed 14 May 2018).
5 *ACTED* (2017), 'Two New Interesting Challenges for ACTED in Azraq Camp, Jordan', 18 October 2017, https://reliefweb.int/report/jordan/two-new-interesting-challenges-acted-azraq-camp-jordan (accessed 14 May 2018).
6 Bhalla, N. (2018), 'Somali Refugees Turn Plastic Waste Warriors in One of the World's Biggest Camps', *Reuters*, 5 June 2018, www.reuters.com/article/us-environment-day-kenya-refugees/somali-refugees-turn-plastic-waste-warriors-in-one-of-worlds-biggest-camps-idUSKCN1J101I (accessed 15 August 2018).
7 Ngala, P. (2018), 'From Waste to Value: Using Solid Waste for Good Sanitation in Uganda's Refugee Settlements', *Reliefweb*, 11 May 2018, https://reliefweb.int/report/uganda/waste-value-using-solid-waste-good-sanitation-uganda-s-refugee-settlements (accessed 14 May 2018).
8 Peters, A. (2018), 'At this Refugee Camp, Residents Are Replanting 40,000 Trees', *Fast Company*, 25 April 2018, www.fastcompany.com/40563628/at-this-refugee-camp-residents-are-replanting-40000-trees (accessed 14 May 2018).
9 Dupire, C. (2017), 'NGO Teaches Refugees to Make A/C Units Out of Plastic Bottles', *The Jordan Times*, 9 September 2017, www.jordantimes.com/news/local/ngo-teaches-refugees-make-ac-units-out-plastic-bottles (accessed 14 May 2018).
10 Cole, R. (2017), 'British Environmental Engineer Urges Action on Waste in Refugee Camps', *Resource*, 12 January 2017, https://resource.co/article/british-environmental-engineer-urges-action-waste-refugee-camps-11605 (accessed 14 May 2018).
11 Cuthbert, O. (2016), 'Broken Glass and Needles: The Waste Pickers Scraping a Living at Jordan's Landfills', *The Guardian*, 27 August 2016, www.theguardian.com/sustainable-business/2016/aug/27/jordan-waste-pickers-syria-refugees-recycling (accessed 14 May 2018).
12 Ryan, A. (2017), 'Growing food from Mattresses: What Experts Can Learn from Working in Refugee Camps', *The Conversation*, 12 December 2017, http://theconversation.com/growing-food-from-mattresses-what-experts-can-learn-from-working-in-refugee-camps-88676 (accessed 14 May 2018).
13 Costa, P. M. (2017), 'The House that Tateh Built . . . Out of Sand-Filled Plastic Bottles', *The Guardian*, 20 June 2017, www.theguardian.com/environment/2017/jun/30/house-tateh-built-sand-filled-recycled-waste-plastic-bottles-western-sahara- (accessed 14 May 2018).
14 Quartier de L'Innovation (2018), 'Combining Social Innovation, Humanitarian Action, and a Circular Economy', 4 May 2018, http://quartierinnovationmontreal.com/en/article/combining-social-innovation-humanitarian-action-and-circular-economy (accessed 15 August 2018).
15 May, M. (2018), 'Smurfit Kappa Turns Its Aid Boxes into Furniture for Edinburgh Direct Aid', *UKFundraising*, 25 April 2018, https://fundraising.co.uk/2018/04/25/smurfit-kappa-turns-aid-boxes-furniture-edinburgh-direct-aid/#.WvmTrFMvzOQ (accessed 14 May 2018).
16 Wainright, P. (2017), 'Why Ikea's Flatpack Refugee Shelter Won Design of the Year', *The Guardian*, 27 May 2017, www.theguardian.com/artanddesign/2017/jan/27/why-ikea-flatpack-refugee-shelter-won-design-of-the-year (accessed 14 May 2018).
17 Meyer, S. (2018), 'Creating a More Sustainable Refugee Camp', *Pacific Standard*, 2 April 2018, https://psmag.com/environment/creating-a-more-sustainble-refugee-camp (accessed 14 May 2018).
18 SDG 6.2 definition: By 2030, achieve access to adequate and equitable sanitation and hygiene for all and end open defecation, paying special attention to the needs of women and girls and those in vulnerable situations.

19 Approach used was based on IDEO.org methodology.
20 Data collected by Sanivation, through continuous monitoring of toilets in Kakuma refugee camp between 2016 and 2018.
21 For EBT testing protocols see http://cleancookstoves.org/technology-and-fuels/testing/protocols.html
22 Traditional Kenyan cook stove.
23 Total cost of fabricating and installing toilets, operating the regular faecal sludge collection service, treating the waste and producing and selling the briquettes, minus the revenues generated from briquette sales.
24 A report carried out by the Boston Consulting Group (BCG) estimates the average lifespan of refugee camps is 17 years. See UNHCR and Boston Consulting Group (2015).
25 Total costs incurred in toilet servicing, waste treatment and briquette production.
26 Calculation: 180,000 (refugees) /5 (average household size) x 365 (days) x 1 (kg waste per HH per day) = 13,140,000 Kgs.
27 For more information see Humanitarian Grand Challenge (2019).

Bibliography

Aol (2018), 'Monsoon fears for Rohingya Refugees in Bangladesh Camps', 27 April.

Baconguis, S. (2003), 'Abandoned Biomass Utilization for Household Energy as a CDM Mechanism for Carbon Dioxide Emission Mitigation', Philippines. http://agris.fao.org/agris-search/search.do?recordID=PH2004001463 (accessed 15 August 2018).

Betts, A., Bloom, L. and Weaver, N. (2015), 'Refugee Innovation: Humanitarian Innovation that Starts with Communities'. www.rsc.ox.ac.uk/files/files-1/refugee-innovation-web-5-3mb.pdf (accessed 15 August 2018).

Bhalla, N. (2018), 'Somali Refugees Turn Plastic Waste Warriors in one of the World's Biggest Camps', *Reuters*, 5 June. www.reuters.com/article/us-environment-day-kenya-refugees/somali-refugees-turn-plastic-waste-warriors-in-one-of-worlds-biggest-camps-idUSKCN1J101I (accessed 15 August 2018).

Birmingham City Council (2018), 'Waste Strategy 2017–2040 Why We Need to Change'. www.birmingham.gov.uk/blog/birmingham-blog/post/11/waste-strategy-2017-2040-why-we-need-to-change (accessed 19 January 2018).

Boston Consulting Group (2015), 'Improving Sanitation in Refugee Camps'. http://wash.unhcr.org/organisation/bcg/ (accessed 15 August 2018).

Clasper, J. (2016), 'Sustainable Waste Management in Lebanon's Refugee Camps', *Virgin Unite*. www.virgin.com/virgin-unite/sustainable-waste-management-lebanons-refugee-camps (accessed 18 March 2018).

Coelho, S. (2010), 'Refugee Camps to use Gas from Human Waste', *Deutsche Welle*, 13 August. www.dw.com/en/refugee-camps-to-use-gas-from-human-waste/a-5908344 (accessed 18 March 2018).

Cogut, A. (2016), 'Open Burning of Waste: A Global Health Disaster'. https://regions20.org/wp-content/uploads/2016/08/OPEN-BURNING-OF-WASTE-A-GLOBAL-HEALTH-DISASTER_R20-Research-Paper_Final_29.05.2017.pdf (accessed 14 May 2018).

Cross, J., Douglas, M., Martin, C., Ray, C., Verhoeven, A. and Grafham, O. (2019), 'Energy and Displacement in Eight Objects: Stories from Sub-Saharan Africa'.

Cross, J. and Murray, D. (2017), 'The Afterlives of Solar Power: Waste and Repair off the Grid in Kenya', *Energy Research & Social Science*, 44 (October), 100–109. doi:10.1016/j.erss.2018.04.034 (accessed 15 August 2018).

de Freytas-Tamura, K. (2017), 'Public Shaming and Even Prison for Plastic Bag Use in Rwanda', *The New York Times*, 28 October. www.nytimes.com/2017/10/28/world/africa/rwanda-plastic-bags-banned.html (accessed 19 January 2019).

Domonoske, C. (2016), 'U.N. Admit Role in Haiti Cholera Outbreak that Has Killed Thousands', *The Two-Way*, 18 April. www.npr.org/sections/thetwo-way/2016/08/18/490468640/u-n-admits-role-in-haiti-cholera-outbreak-that-has-killed-thousands (accessed 14 May 2018).

Edwards, A. (2016), 'South Sudan Fighting Drives Surge of Refugees to Uganda', *The UN Refugee Agency*. www.unhcr.org/uk/news/latest/2016/7/57973cde4/south-sudan-fighting-drives-surge-refugees-uganda.html (accessed 14 May 2018).

Foote, A. M., Woods, E., Fredes, F. and Leon, J. S. (2017), 'Rendering Faecal Waste Safe for Reuse Via a Cost-Effective Solar Concentrator', *Journal of Water Sanitation and Hygiene for Development*, 7 (2), 252–259. doi:10.2166/washdev.2017.112 (accessed 15 August 2019).

Grafham, O. and Lahn, G. (2018), 'The Costs of Fuelling Humanitarian Aid', *The Royal Institute of International Affairs*. www.chathamhouse.org/sites/default/files/publications/research/2018-12-10-Costs-Humanitarian-Aid2.pdf (accessed 6 March 2019).

Grand Challenges Canada (2019), 'Creating Hope in Conflict: A Humanitarian Grand Challenge', *Grand Challenges Canada*, Canada.

IDEO (2018), 'Human-Centered Design Sits at the Intersection of Empathy and Creativity'. www.ideo.org/approach (accessed 15 August 2018).

Magalini, F., Shinha-Khetriwal, D., Rochat, D., Huismann, J., Munyamba, S., Oliech, J., Nnorom, I. C. and Mbera, O. (2016), 'Electronic Waste (e-waste) Impacts and Mitigation Options in the Off-Grid Renewable Energy Sector'. www.gov.uk/dfid-research-outputs/electronic-waste-e-waste-impacts-and-mitigation-options-in-the-off-grid-renewable-energy-sector (accessed 15 May 2018).

Mohammad, A. and Kenneth, P. (2016), 'Waste Management in Zaatari Refugee Camp/Mafraq-Jordan', 10th International Conference of Establishment of Cooperation Between Companies and Institutions in the Nordic Countries, the Baltic Region and the World. https://lup.lub.lu.se/search/publication/0edc01b7-031e-43dc-b916-e19b87cd121c (accessed 14 May 2018).

Nyoka, R., Foote, A. D., Woods, E., Lokey, H., O'Reilly, C., Magumba, F., Okello, O., Mintz, E. D., Marano, N. and Morris, J. F. (2017), 'Sanitation Practices and Perceptions in Kakuma Refugee Camp, Kenya: Comparing the Status Quo with a Novel Service-Based Approach', *PLoS One*, 12. doi:10.1371/journal.pone.0180864 (accessed 15 August 2018).

Owiiti, J. (2003), 'Evaluation of Solar Cooking project in Kakuma Refugee Camp', *Center for Independent Research and Energy for Sustainable Development Africa*.

Price, R. (2017), 'Environmental Impact Assessments in Refugee Crises', *Knowledge, Evidence and Learning for Development*. https://assets.publishing.service.gov.uk/media/5a5f2cdbed915d7dfea66cd8/229-EIA-in-refugee-crises.pdf (accessed 14 May 2018).

Sanivation (2018), www.saivation.com (accessed 15 August 2018).

Sukru, T. (2017), 'Kenya 2017 Key Response Figures'. https://data2.unhcr.org/en/documents/details/63607 (accessed 15 August 2018).

UNHCR (2018a), 'Kenya Registered Refugees and Asylum-Seekers'. www.unhcr.org/ke/wp-content/uploads/sites/2/2018/03/Kenya-Infographics-February-2018.pdf (accessed 15 August 2018).

UNHCR (2018b), 'UNHCR WASH Standards and Indicators – March 2018'. http://wash.unhcr.org/unhcr-wash-standards-and-indicators-for-refugee-settings/ (accessed 10 February 2019).

UNHCR and Boston Consulting Group (2015), 'Improving Sanitation in Refugee Camps'. http://wash.unhcr.org/organisation/bcg/ (accessed 15 August 2018).

Watts, J. (2018), 'Eight Months on, Is the World's Most Drastic Plastic Bag Ban Working?' *The Guardian*, 25 April. www.theguardian.com/world/2018/apr/25/nairobi clean-up-highs-lows-kenyas-plastic-bag-ban (accessed19 January 2019).

World Bank (2017), 'PPP Reference Guide 3.0'. https://ppp.worldbank.org/public-private-partnership/library/ppp-reference-guide-3-0 (accessed 15 August 2018).

Index

Note: Page numbers in *italics* indicate figures and page numbers in **bold** indicate tables.

www.ingramcontent.com/pod-product-compliance
Ingram Content Group UK Ltd.
Pitfield, Milton Keynes, MK11 3LW, UK
UKHW011011180926
475701UK00004B/686
* 9 7 8 1 0 3 2 0 8 2 1 3 4 *